GUIDE TO MATHEMATICAL METHODS

This book is the property
of Royston Charles Cheater.

Consultant Editor: David A. Towers,
Senior Lecturer in Mathematics,
University of Lancaster

Titles available

Abstract Algebra
Analysis
Linear Algebra
Mathematical Models

Further titles are in preparation

Guide to Mathematical Methods

John Gilbert

*Department of Mathematics, University of Lancaster,
Bailrigg, Lancaster LA1 4YL*

MACMILLAN

First published 1991 by
THE MACMILLAN PRESS LTD
Houndmills, Basingstoke, Hampshire RG21 2XS
and London
Companies and representatives
throughout the world

ISBN 0–333–49209–9

A catalogue record for this book is available
from the British Library

Printed in Hong Kong

Reprinted (with corrections) 1992

CONTENTS

Consultant Editor's Foreword   viii

Preface   ix

Symbols, Notation and Greek Letters   xi

1   FUNCTIONS   **1**
   1.1   Sets and intervals   1
   1.2   Functions   4
   1.3   Limits and continuity   6
   1.4   Polynomials   8
   1.5   Rational functions   9
   1.6   Trigonometric functions   9
   1.7   Exponential and logarithmic functions   11
   1.8   Hyperbolic functions   14
   1.9   Composite and inverse functions   15
   1.10   Answers to exercises   21

2   DIFFERENTIATION   **26**
   2.1   Introduction   26
   2.2   Derivatives of combinations of functions   29
   2.3   Derivatives of trigonometric functions   33
   2.4   Derivatives of exponential and logarithmic functions   36
   2.5   Higher derivatives   38
   2.6   Power series expansions   40
   2.7   List of standard derivatives   42
   2.8   Answers to exercises   44

3   INTEGRATION   **48**
   3.1   Area and definite integrals   48
   3.2   Speed and distance, force and work   51

3.3	The Fundamental Theorem of Calculus	52
3.4	Standard integrals and properties of integrals	54
3.5	Integration by substitution	55
3.6	Integrals involving the factor $\sqrt{\pm a^2 \pm x^2}$	59
3.7	Integration by parts	64
3.8	Partial fractions	69
3.9	Systematic integration of rational functions	73
3.10	Rational trigonometric functions	77
3.11	Improper integrals	78
3.12	List of standard integrals	81
3.13	Answers to exercises	84
4	**LINEAR EQUATIONS, DETERMINANTS AND MATRICES**	**96**
4.1	Introduction	96
4.2	Determinants	100
4.3	Three or more simultaneous equations	103
4.4	Matrices	106
4.5	Square matrices	112
4.6	Numerical solution of linear equations	121
4.7	Answers to exercises	126
5	**VECTORS**	**131**
5.1	Introduction	131
5.2	Coordinate systems	131
5.3	The algebra of vectors	134
5.4	Unit vectors and direction cosines	137
5.5	Scalar products	140
5.6	Vector products	142
5.7	Triple products	146
5.8	Lines and planes	149
5.9	Vector equations of curves in space	156
5.10	Differentiation of vector functions	159
5.11	Arclength	162
5.12	Answers to exercises	166
6	**FUNCTIONS OF TWO VARIABLES**	**175**
6.1	Introduction	175
6.2	The standard family of functions	176
6.3	Graphical representation	176
6.4	Functions of three or more variables	181
6.5	Partial derivatives	181
6.6	Chain rules	183
6.7	Directional derivatives	185

6.8	Higher partial derivatives	186
6.9	Maxima and minima	190
6.10	Answers to exercises	195

7	**LINE INTEGRALS AND DOUBLE INTEGRALS**	**202**
7.1	Vector fields	202
7.2	Line integrals	203
7.3	Properties of line integrals	206
7.4	Conservative fields	209
7.5	Double integrals	215
7.6	Change of variables	226
7.7	Green's theorem	234
7.8	Answers to exercises	238

8	**COMPLEX NUMBERS**	**247**
8.1	Introduction	247
8.2	The algebra of complex numbers	248
8.3	Solution of equations	253
8.4	Equalities and inequalities	255
8.5	Polar form of complex numbers	257
8.6	Exponential form of complex numbers	261
8.7	Trigonometric identities	263
8.8	Answers to exercises	265

9	**DIFFERENTIAL EQUATIONS**	**270**
9.1	Introduction	270
9.2	Differential equations of Type 1 (separable)	275
9.3	Differential equations of Type 2	280
9.4	Differential equations of Type 3 (exact)	282
9.5	Differential equations of Type 4 (linear)	283
9.6	An alternative way of solving first-order linear differential equations	287
9.7	Solution of second-order differential equations	291
9.8	Complementary function of second-order differential equations	292
9.9	General solution of non-homogeneous equation	295
9.10	Initial and boundary conditions	299
9.11	Answers to exercises	302

Index		**309**

CONSULTANT EDITOR'S FOREWORD

Wide concern has been expressed in tertiary education about the difficulties experienced by students during their first year of an undergraduate course containing a substantial component of mathematics. These difficulties have a number of underlying causes, including the change of emphasis from an algorithmic approach at school to a more rigorous and abstract approach in undergraduate studies, the greater expectation of independent study, and the increased pace at which material is presented. The books in this series are intended to be sensitive to these problems.

Each book is a carefully selected, short, introductory text on a key area of the first-year syllabus; the areas are complementary and largely self-contained. Throughout, the pace of development is gentle, sympathetic and carefully motivated. Clear and detailed explanations are provided, and important concepts and results are stressed.

As mathematics is a practical subject which is best learned by doing it, rather than watching or reading about someone else doing it, a particular effort has been made to include a plentiful supply of worked examples, together with appropriate exercises, ranging in difficulty from the straightforward to the challenging.

When one goes fellwalking, the most breathtaking views require some expenditure of effort in order to gain access to them: nevertheless, the peak is more likely to be reached if a gentle and interesting route is chosen. The mathematical peaks attainable in these books are every bit as exhilarating, the paths are as gentle as we could find, and the interest and expectation are maintained throughout to prevent the spirits from flagging on the journey.

Lancaster, 1990

David A. Towers
Consultant Editor

PREFACE

Mathematical methods have been used extensively through the years to solve problems arising in science and engineering, while more recently they have become more and more used by social scientists. Consequently there is a need for students in these disciplines to acquire a practical working knowledge of mathematics. This book is intended to enable such students, at about first-year undergraduate level, to acquire the mathematical skills required to solve problems arising in their subject, without becoming unnecessarily embroiled in the finer mathematical details. Even though the presentation is not mathematically rigorous, it should also provide a useful supplement to methods courses for mathematics students.

There is always likely to be a conflict in any book on mathematics between the need to explain as simply as possible how an idea or method works and the need for rigour. I have taken the line that the former is more important for students who are more interested in applications than the mathematics itself. Thus, very few proofs are given, and although I have attempted to stay fairly honest, I have glossed over some of the more analytic details, especially those concerned with limiting processes. Hopefully, anyone who also studies *Analysis* should have no difficulty in recognising the stage at which I have fallen short of full rigour, and accept that this has to be so in a book of this type. For a user with no need to question or understand the method itself, one could present a set of recipes to cover all or most of the standard problems. The drawback of this, of course, is when a non-standard problem arises, which calls for a modification of the method. I have therefore devoted a fair amount of space to the mathematical ideas behind the methods; students who wish to use the text just as a recipe book may avoid the mathematics, but those who wish to be able to apply the methods to a wider set of problems will wish to study the mathematical parts in more detail.

A difficult decision has to be made in planning a methods book as to whether applied problems should be studied in the context of the various user disciplines. Interesting and instructive though this may be, however, it can often distract attention from the mathematical methods themselves, and would be likely to lengthen the text unacceptably. I have therefore, on the whole, avoided involvement with such problems, except where the initial motivation or understanding of the mathematics has called for it. However, I have included at the end of each chapter a number of applied exercises; although these are couched in the language of the applications, I have omitted the derivation, and simply stated the mathematical equations to be solved. Not all chapters are concerned with topics which lead directly to applied problems, and in such cases I have used the exercises to extend the mathematical ideas given in the chapter.

The style and pace of the book is intended to be such that an average student should be able to work through it with the absolute minimum of external assistance. To help consolidate the material as soon as possible, exercises are given at the end of each section. New concepts are introduced in a common-sense fashion, starting with examples from applications to motivate the formal mathematical definition. The derivation of methods is often done at first for a simple case, in order to give an intuitive understanding. A more general derivation usually follows, unless this involves complication or deep mathematics, in which case the method is induced from the simple case. Topics, such as set theory, are dealt with superficially when applications require only a general appreciation of their results. Other topics are dealt with mostly in a conventional way, but linear equations, apart from being important in their own right, are used as a vehicle to introduce determinants and matrices in a very natural way. Vectors are, somewhat unusually, firmly based in coordinate systems, in the belief that any vector calculation will ultimately have to be translated into coordinate form to apply the results to a practical situation.

Most of this book is based very closely on notes provided for Sheffield undergraduates to learn from at their own pace, and I am greatly indebted to Keith Austin and Sheffield University colleagues John Baker, David Chellone, David Jordan, Chris Knight, David Sharpe, Peter Vámos (now at Exeter) and Roger Webster for providing so much of the source material. Chapters 4, 7 (the part on double integrals) and 9 are based on my own experience of teaching service mathematics at Lancaster over a number of years, I must also record my thanks to Keith for the many very constructive suggestions that he made, almost all of which I gratefully followed. Finally, I could not miss this opportunity of thanking my wife for bearing so nobly the considerable time I spent shut off from her in order to complete the book.

Lancaster, 1990

J.G.

SYMBOLS, NOTATION AND GREEK LETTERS

SYMBOLS

$\mathbb{N}$	Set of all natural numbers
$\mathbb{Z}$	Set of all integers
$\mathbb{Q}$	Set of all rational numbers
$\mathbb{R}$	Set of all real numbers
$\mathbb{R}^+$	Set of all positive real numbers
$\mathbb{C}$	Set of all complex numbers
$\in$	Belongs to
$\subset$	Is contained in
$\cap$	Intersection
$\cup$	Union
$\setminus$	Excluding
$\varnothing$	The empty set
∞	Infinity
$>$	Greater than
$\geq$	Greater than or equal to
$<$	Less than
$\leq$	Less than or equal to
$\approx$	Approximately equal to
$\rightarrow$	Tends to
$\neq$	Not equal to
$\triangle$	Discriminant

NOTATION

$\{a, b, \ldots\}$	Set whose elements are $a, b, \ldots,$
$[a, b]$	Closed interval (set of all real numbers x satisfying $a \leq x \leq b$)

(a, b)	Open interval (set of all real numbers x satisfying $a < x < b$)
$f : A \to B$	Function mapping set A into set B
$\lvert x \rvert$	Modulus of the real number x
$[x]$	Largest integer not greater than x
$x \to a_-$	x tends to a from values less than a
$x \to a_+$	x tends to a from values greater than a
$\lim\limits_{x \to a}$	Limit as x tends to a
$\exp$	Exponential function
$\exp(x) = e^x$	Value of the exponential function at x
$\log_a$	Logarithmic function to base a
$\log_e x = \ln x$	Value of the logarithm of x to base e
$(f \circ g)$	The composition of the functions g and f
$(f \circ g)(x) = f(g(x))$	Value of $(f \circ g)$ at x
f^{-1}	Inverse of the function f
$\dfrac{dy}{dx}$	Derivative of y with respect to x
$f'(x)$	Derivative of the function f at x
$\dfrac{d}{dx}$	Differential operator
δx	Small change in x
$\sum\limits_{i=1}^{n} a_i$	Sum of $a_1, a_2, \ldots, a_n$
$\int_a^b f(x)\,dx$	Definite integral of f from a to b
$\int f(x)\,dx$	Indefinite integral of f
$[p(x)]_a^b$	Has the value $p(b) - p(a)$
$\begin{vmatrix} a & b \\ c & d \end{vmatrix}$	Determinant with value $ad - bc$
$\begin{bmatrix} a & b \\ c & d \end{bmatrix}$	Matrix with two rows and two columns
a_{ij}	Element of matrix in the ith row and jth column
$\overrightarrow{PQ}$	Vector
$\mathbf{r}$	Vector
$(\mathbf{r})_k$	kth component of $\mathbf{r}$
(x, y, z)	Position vector with components x, y, z
$\lvert \mathbf{r} \rvert$	Magnitude (length) of $\mathbf{r}$

$\hat{\mathbf{r}}$	Unit vector in the direction of $\mathbf{r}$
$\mathbf{i, j, k}$	Unit vectors in the directions of the coordinate axes
l, m, n	Direction cosines
$\mathbf{r \cdot s}$	Scalar or dot product
$\mathbf{r \times s}$	Vector or cross product
$\dfrac{dy}{dx}\Big|_{t=t_0}$	Value of $\dfrac{dy}{dx}$ at $t = t_0$
$\dfrac{\partial f}{\partial x} = f_x$	Partial derivative of f with respect to x
$\dfrac{\partial f}{\partial y} = f_y$	Partial derivative of f with respect to y
$\dfrac{\partial^2 f}{\partial x^2} = f_{xx}$	Second partial derivative
$\dfrac{\partial^2 f}{\partial x \partial y} = f_{yx}$	Second partial derivative
$\nabla f = \dfrac{\partial f}{\partial x}, \dfrac{\partial f}{\partial y}, \dfrac{\partial f}{\partial z}$	Gradient of f
$\mathbf{V}$	Vector field
$\displaystyle\int_C \mathbf{V} \cdot d\mathbf{r}$	Line integral along the curve C given by $\mathbf{r}$
$\displaystyle\int_C P(x, y)\, dx + Q(x, y)\, dy$	Line integral of $P(x, y)\mathbf{i} + Q(x, y)\mathbf{j}$ along C given by $\mathbf{r}(s) = x(s)\mathbf{i} + y(s)\mathbf{j}$
$\displaystyle\oint_C \mathbf{V} \cdot d\mathbf{r}$	Line integral round closed curve C
$\displaystyle\iint_R f\, dA$	Double integral over a region R
$\displaystyle\iint_R f(x, y)\, dx\, dy$	Double integral over a region R of the xy plane
$\displaystyle\int_a^b \int_{g(x)}^{h(x)} f(x, y)\, dy\, dx$	Repeated integral
$\dfrac{\partial(x, y)}{\partial(u, v)}$	Jacobian
$z = x + iy$	Complex number
$\mathrm{Re}(z) = x$	Real part of $z = x + iy$

$\text{Im}(z) = y$	Imaginary part of $z = x + iy$
$\lvert z \rvert$	Modulus of the complex number z
$\bar{z} = x - iy$	Complex conjugate of $z = x + iy$
$r(\cos\theta + i\sin\theta)$	Polar form of a complex number
$re^{i\theta}$	Exponential form of a complex number

GREEK LETTERS

α	alpha
β	beta
γ	gamma
δ	delta
ε	epsilon
θ	theta
κ	kappa
λ	lambda
μ	mu
ν	nu
π	pi
ρ	rho
σ	sigma
Σ	Sigma
ϕ	phi
ω	omega
Ω	Omega

1 FUNCTIONS

1.1 SETS AND INTERVALS

In order to define functions, which are the main concern of this chapter, we first need to know a little about set notation. Although we shall start with general sets, we shall move on rapidly to special kinds of sets, called intervals, which are essential for the study of functions.

The word 'set' is the collective noun of mathematics. Whereas the farmer would refer to a flock of sheep and the admiral to a fleet of ships, the mathematician would refer to a set of sheep and a set of ships.

Notation

A *set* is a collection of objects, called elements. In the example above, an element in the set of sheep is a sheep. We usually use a capital letter to denote a set. If x is an element of the set A, we write

$$x \in A \text{ (read as `} x \text{ belongs to } A\text{')}$$

We specify the elements, $a_1, a_2, \ldots, a_n$, of a set A by writing

$$A = \{a_1, a_2, \ldots, a_n\}$$

Whether used in the context of sets or not, the three dots between the commas always stand for a sequence of objects of the kind implied by the first few and the last. If no last object is specified, the sequence is assumed not to terminate—that is, it is infinite. This is occasionally used backwards, as in the specification of $\mathbb{Z}$ below.

It is often necessary to exclude a certain element or elements from a general set; for example, if A is the set above, but we wish to exclude the elements a_3 and a_7, we write

$$A \setminus \{a_3, a_7\} \text{ (read as `} A \text{ not } a_3 \text{ or } a_7\text{')}$$

1

There are standard symbols for a number of useful sets:

$\mathbb{N} = \{1, 2, 3, \ldots\}$

$\mathbb{Z} = \{\ldots, -3, -2, -1, 0, 1, 2, 3, \ldots\}$

$\mathbb{Q}$ = the set of rational numbers or fractions (e.g. $\frac{1}{2}$, $-\frac{3}{2}$)

$\mathbb{R}$ = the set of real numbers (e.g. e, π, $\sqrt{2}$)

$\mathbb{C}$ = the set of complex numbers (see Chapter 8)

We say that a set B is a *subset* of a set A if every element in B is also in A and write this as $B \subset A$. We also say in this case that B is contained in A, or A contains B.

The sets above satisfy

$$\mathbb{N} \subset \mathbb{Z} \subset \mathbb{Q} \subset \mathbb{R} \subset \mathbb{C}$$

Each bigger set arose historically as an extension of the smaller set to include solutions to a wider class of problems.

We might also be interested in combinations of sets. Suppose we have two sets A and B which are represented in Figure 1.1(a) by the interior of the two curves labelled A and B. Then:

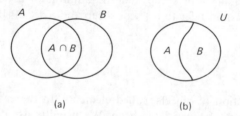

(a)   (b)

Fig. 1.1

Definitions

The *intersection* of A and B, denoted by $A \cap B$, is the set of elements which belong to both A and B.

The *union* of A and B, written $A \cup B$, is the set of elements which are in A or B or both.

If the sets A and B have no element in common, we say that they do not intersect, and write this mathematically as

$$A \cap B = \varnothing$$

where $\varnothing$ is the *empty set*—that is, the set containing no elements.

It may be that $A \cup B$ gives the whole of the set we are considering; we write this as

$$A \cup B = U$$

where U is the *universal set*, or the set of all elements. Unlike $\varnothing$, U is context-dependent, so we must always define what we mean by it.

If the sets A and B are such that $A \cap B = \varnothing$ and $A \cup B = U$, then we say that A is the *complement* of B (or B is the complement of A), writing $A = \backslash B$ (or $B = \backslash A$). An example of this situation is shown in Figure 1.1(b). This is another use of the exclusion symbol introduced earlier; we can think of $\backslash B$ as being a shorthand version of $U \backslash B$.

Example 1.1.1

Let $U = \mathbb{N} = \{1, 2, 3, \ldots\}$, and let $A = \{2, 4, 6, \ldots\}$, the set of even natural numbers, and $B = \{1, 3, 5, \ldots\}$, the set of odd natural numbers. Then $A \cap B = \varnothing$ and $A \cup B = U = \mathbb{N}$. Thus $B = \backslash A$.

There is a whole area of mathematics devoted to the study of sets and their properties, but we shall leave this and concentrate on the set $\mathbb{R}$ and its subsets, which we shall find extremely useful in our study of methods. $\mathbb{R}$ itself is often referred to as the *real line*, because of its geometric connection with lengths, such as 2π, the circumference of the unit circle, or $\sqrt{2}$, the length of the third side of a right-angled triangle whose other two sides are of unit length. $\mathbb{R}$ is the set which is needed in all the practical calculations involving physical dimensions.

It is frequently the case that we wish to include only a part of the real line in our considerations; we achieve this by defining the following subsets of $\mathbb{R}$.

Definitions

The *closed interval*, $[a, b]$, is the set of values of x which satisfy the inequality $a \leq x \leq b$. (Since only real numbers can satisfy this inequality, the interval $[a, b]$ consists of real numbers.)

The *open interval*, (a, b), is the set of values of x which satisfy the inequality $a < x < b$. (An equivalent but less-used notation for an open interval is $]a, b[$.)

The *length* of both these intervals is $b - a$.

We can actually mix the two kinds of bracket—for example, $[-1, 3)$ is the set of values of x for which $-1 \leq x < 3$. Thus, a square bracket means that the end-point of the interval is included, while a round bracket means that the end-point is excluded. Because ∞ (see Example 1.3.1.2 for the meaning of this symbol) is not a specific number, we use a round bracket with it. For example, $(-\infty, 0]$ is the set of values of x for which $x \leq 0$, while $(-\infty, \infty)$ is the whole of $\mathbb{R}$.

EXERCISE 1.1.1

Use interval notation to describe the sets of values of x which satisfy the inequalities.

(i) $1 \leq x < 3$; (ii) $x \leq -5$; (iii) $-2 < x \leq 1$.

1.2 FUNCTIONS

Probably the single most important concept in mathematics is that of a function. It occurs in almost every branch of mathematics and in many applications, so it is not surprising that it is a concept of great generality. However, for our purpose it will be sufficient to give a fairly restricted definition. Before doing even this, we shall try to give an intuitive idea of a function with the help of some commonly occurring examples, so that when we do give the formal definition, it will make better sense.

Examples 1.2.1

1. Suppose that an object is dropped from a height of 20 m above the ground. Its height in metres after t seconds, assuming no air resistance, is given by $h = 20 - \frac{1}{2}gt^2$, where g is the acceleration due to gravity in metres per second squared. Of course this formula only makes sense for values of t between 0 and the time, $\sqrt{40/g}$ seconds, at which the object hits the ground. We say that h *is a function of* t. The behaviour of the height, h, is determined simply by the value of the time, t.
2. Let A be the area of a sector of the unit circle (that is, the circle of radius 1) of angle t. Then $A = \frac{1}{2}t$. (We can obtain this by noting that the area of the sector is simply a fraction $t/2\pi$ of the area of the whole circle, π.) Here A is a function of t. The area of the sector is determined by the size of its angle.
3. Let T be the temperature in degree C at Manchester Airport on the day whose date is t. In the first two examples, h and A are functions which are expressed by formulae; in this example no formula is known, but, for whatever time we specify, there is a corresponding value of T. We therefore still say that T is a function of t.

We make some simple observations about these examples:

(i) We have used the symbol t to represent different quantities. This is just a matter of convenience; for example, if we used x for the angle of the sector in Example 1, the area would be $A = \frac{1}{2}x$ and A would be a function of x.

(ii) It is clear that each value of t gives only one answer. We say that the answer is *unique*.

(iii) In the first example, it was important that we only used the formula for certain values of t. Example 3 only makes sense if we specify integer values for t, since t represents a date. These types of restriction are often important in more general examples.

We are now ready to give the definition of a function.

Definition

A *function* $f: A \to B$ is a rule which associates with each member x of the set A a unique member y of the set B. The set A is called the *domain* of f and we say that f is defined on A. We write $y = f(x)$ and call $f(x)$ the *value* of f at x. The set of values of f as x takes all values in A is called the *range* of f.

Although we have defined a function in terms of general sets A and B, these will both usually consist of $\mathbb{R}$ or a subset of $\mathbb{R}$. We read $f: A \to B$ as 'f maps A into B'; if it is clear what the sets A and B are from the context, we shall omit reference to them. A function is defined by the values it takes at all points of the set A.

It may help to think of a function as a 'number-vending machine'. If you push in a number, out comes at most *one* other number. For example, consider the function defined by $f(x) = 1/x^2$. If you push in the number 2, out comes the number $\frac{1}{4}$; if you put in $\frac{1}{3}$, out comes 9. What happens if you put in the number 0? Since we cannot divide by zero, we should not expect an answer. We say that the function is *not defined* at $x = 0$. In fact, this is a case where we should have given the domain of f as part of its definition; thus we should have said

'the function $f: \mathbb{R} \setminus \{0\} \to \mathbb{R}$ is defined by $f(x) = 1/x^2$'

or

'the function f is defined by $f(x) = 1/x^2$, $x \neq 0$'

Here $\mathbb{R} \setminus \{0\}$ means the set of real numbers excluding zero.

Consider now the rule $f(x) = y$ where $y^2 = x$. For $x = 4$ we have $f(x) = 2$ or -2, so there are two possible values. We therefore cannot use this rule for defining a function. We could use it, however, if we insisted on always taking the positive square root, for example.

Examples 1.2.2

1. The function $h: [0, \sqrt{40/g}] \to \mathbb{R}$ is defined by $h(t) = 20 - \frac{1}{2}gt^2$. This should be recognised as the first of Examples 1.2.1 in terms of our new notation. Note that the set of possible values of t, the domain of h, is included in

the definition as $[0, \sqrt{40/g}]$. For each value of the time, t, $h(t)$ is the height of the object. The range of h is $[0, 20]$.

2. For the second of Examples 1.2.1, we can define the function A by $A(x) = \frac{1}{2}x$. Then $A(x)$ is the area of a sector of the unit circle of angle x. In this case we have not bothered to specify the set of permissible values of x, but the only sensible values are in the interval $[0, 2\pi]$, when the range is $[0, \pi]$.

3. For the third of Examples 1.2.1, we can only define the function $T: A \rightarrow R$, where the domain A of T is the set of possible dates, by giving a table of values $T(t)$ for every t in A.

As we have seen, many functions can be defined by a formula; such a formula can often be expressed in terms of a few *basic* functions, which we shall now describe. Later we shall combine these basic functions in various ways to generate a whole collection of other functions, which we shall call *the family of standard functions*.

EXERCISES 1.2.1

1 Which of the following rules define y as a function of x? Give reasons for your answers. (You may need to refer to later sections in this chapter if you have not already met tan and e.)
  (i) For each number x, y is given by $\tan y = x$.
  (ii) For each positive number x, y is given by $e^y = x$.
  (iii) For each number x, y satisfies $y^2 = x^2 + 1$.
  (iv) For each number x, $y = |x|$ (this is the modulus function defined by $y = x$ if $x \geqslant 0$, $y = -x$ if $x < 0$).
  (v) For each number x, $y = [x]$ (that is, the largest integer not greater than x. For example, $[2 \cdot 6] = 2$, $[-3 \cdot 4] = -4$).

2 Give domains as large as possible for the functions given by:
  (i) $y = \sqrt{x}$; (ii) $y = 1/(x-2)$; (iii) $y = \sqrt{9 - x^2}$.

3 Give the ranges of the following functions:
  (i) $y = 2 \cos x - 3$; (ii) $y = \log x (x > 0)$;
  (iii) $y = \sqrt{1 - x^2}(-1 \leqslant x \leqslant 1)$; (iv) $y = |x|$; (v) $y = [x]$.

1.3 LIMITS AND CONTINUITY

Figure 1.2 shows three types of function we might encounter; that in (a) has no breaks in it, while each of those in (b) and (c) has an obvious break or *discontinuity* at $x = a$.

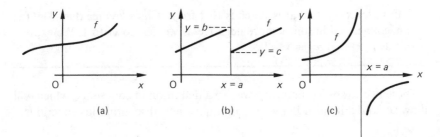

Fig. 1.2

Let us see how we should define the function in (*b*) at $x = a$. We cannot give it two values, since the value must be unique. To get round this, we set $f(a) = c$, say, and consider the value of $f(x)$ as x increases towards the value a (but never quite reaches it!). We express this mathematically as $x \to a_-$, read as 'x tends to a from values less than a'. We now say that $f(x) \to b$ as $x \to a_-$; this means that the closer that x comes to a (from values less than a) the closer $f(x)$ comes to b. b is called the limit of $f(x)$ as $x \to a_-$. We also express this by the statement

$$\lim_{x \to a_-} f(x) = b$$

In graphical terms, we could depict this by leaving a gap in the graph just to the left of $x = a$, as shown in Figure 1.2(b).

Examples 1.3.1

1. Define the function f by

$$f(x) = \begin{cases} x + 1 & x < 2 \\ x & x \geq 2 \end{cases}$$

Note that the value of f at $x = 2$ comes from the second part of the definition as 2, while from the first part $f(x) \to 3$ as $x \to 2_-$. This is the function shown in Figure 1.2(b), with $a = 2$, $b = 3$, $c = 2$.

2. Define the function f by

$$f(x) = \begin{cases} 1/(1 - x) & x \neq 1 \\ 0 & x = 1 \end{cases}$$

What happens to f as $x \to 1_-$? Since $x < 1$, $f(x)$ is positive and becomes larger and larger as x gets closer to 1. We express this by writing $f(x) \to \infty$ ($f(x)$ tends to infinity) as $x \to 1_-$. We should think of ∞, not as one particular large number, but as being larger than any positive number we can specify.

7

If we let $x \to 1_+$ (that is, x tends to 1 from values greater than 1), $f(x)$ is negative, and becomes more negative as we get closer to 1. We express this as $f(x) \to -\infty$ as $x \to 1_+$.

These two examples point the way to a definition of continuity, which will allow us to distinguish between continuous and discontinuous functions.

Definitions

The function f is *continuous* at $x = a$ if

$$\lim_{x \to a_-} f(x) = f(a) = \lim_{x \to a_+} f(x)$$

The function is continuous on an interval if it is continuous at every point of the interval.

A function which is not continuous is discontinuous.

The student should check from the definition that a function such as that shown in Figure 1.2(a) is continuous.

1.4 POLYNOMIALS

The general polynomial of degree n in x may be written as

$$y = f(x) = a_n x^n + a_{n-1} x^{n-1} + \cdots + a_1 x + a_0$$

where $a_0, a_1, \ldots, a_n$ are numbers and $a_n \neq 0$. The *degree* of a polynomial is simply the highest power of x occurring. Polynomials are probably the most useful functions in mathematics; this is because they are so easy to manipulate and compute the values of. Examples of degree 1 and 3, respectively, are given by $f(x) = 5x + 1$ and $f(x) = 2x^3 - 3x + 4$. For polynomials of the form $f(x) = x^n$, the graphs take the characteristic form shown in Figure 1.3 when $n = 1$, n is even or n is odd.

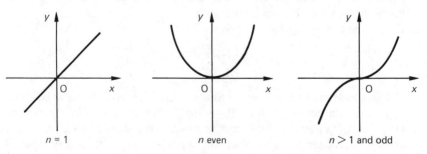

$n = 1$ $\qquad$ n even $\qquad$ $n > 1$ and odd

Fig. 1.3 Graphs of $y = x^n$

1.5 RATIONAL FUNCTIONS

The general rational function, f, is defined by

$$f(x) = \frac{p(x)}{q(x)}$$

where p and q are both polynomials. We must be careful to specify in the definition of f a domain which excludes values of x at which $q(x)$ is zero. Examples of rational functions are

$$f : \mathbb{R} \setminus \{0\} \to \mathbb{R} \text{ defined by } f(x) = \frac{1}{x}$$

$$g : \mathbb{R} \setminus \{1\} \to \mathbb{R} \text{ defined by } g(x) = \frac{2x^2 - 3x + 4}{x^3 - 1}$$

Here the domain of f is the set of real numbers excluding zero; the range of f is $\mathbb{R} \setminus \{0\}$. The domain of g is $\mathbb{R}$ excluding the value 1, and its range is $\mathbb{R} \setminus \{0\}$, although this is a little harder to see.

1.6 TRIGONOMETRIC FUNCTIONS

Trigonometric functions are functions of angles which are measured in radians. In the unit circle shown in Figure 1.4(a), the angle x equals the length of the arc IP, since OP = 1. PN is perpendicular to OI. Then the sine, cosine and tangent functions are defined by

$$\sin x = \text{NP}, \cos x = \text{ON} \quad \text{and} \quad \tan x = \sin x / \cos x = \text{NP/ON}$$

The reciprocals of these functions are given separate names for convenience.

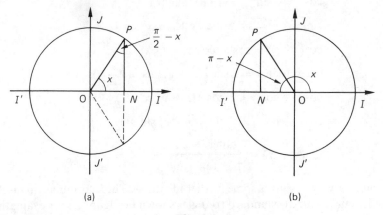

(a)   (b)

Fig. 1.4

Thus,

$$\operatorname{cosec} x = 1/\sin x, \ \sec x = 1/\cos x \quad \text{and} \quad \cot x = 1/\tan x$$

In Figure 1.4(a) the angle x is in the first quadrant—that is, $0 \leq x \leq \pi/2$.

What happens when $x > \pi/2$, as in Figure 1.4(b)? The definitions are unchanged, but we must note that when P is above the line I'I, NP is positive, while NP is negative when P is below I'I. Similarly, ON is positive when P is to the right of the line J'J and negative when it is to the left. The definitions above then apply to all angles, including negative ones.

We deduce quite easily a number of relations:

$$\sin(-x) = -\sin x; \quad \cos(-x) = \cos x; \quad \sin\left(\frac{\pi}{2} - x\right) = \cos x;$$

$$\cos\left(\frac{\pi}{2} - x\right) = \sin x$$

(all from Figure 1.4(a));

$$\sin(\pi - x) = \sin x; \ \cos(\pi - x) = -\cos x \ \text{(from Figure 1.4b)}.$$

If we take $x = 0$, PN has zero length and ON = OP = 1; the definitions then yield

$$\sin 0 = \tan 0 = 0; \ \cos 0 = 1$$

Taking $x = \pi/2$, ON has length 0 and PN = OP = 1, so that

$$\sin(\pi/2) = 1; \ \cos(\pi/2) = 0; \ \tan(\pi/2) = \infty$$

Note that $\tan x$ actually becomes larger than any positive number as $x \to \pi/2$, but it is convenient to write it as equal to ∞.

Other useful identities are

$$\cos^2 x + \sin^2 x = 1 \ \text{(Pythagoras)};$$

$$\sin(x + y) = \sin x \cdot \cos y + \cos x \cdot \sin y;$$

$$\cos(x + y) = \cos x \cdot \cos y - \sin x \cdot \sin y;$$

$$\sin x + \sin y = 2 \sin \tfrac{1}{2}(x + y) \cos \tfrac{1}{2}(x - y);$$

$$\cos x + \cos y = 2 \cos \tfrac{1}{2}(x + y) \cos \tfrac{1}{2}(x - y);$$

$$\cos x - \cos y = 2 \sin \tfrac{1}{2}(x + y) \sin \tfrac{1}{2}(y - x);$$

$$\tan(x + y) = \frac{\tan x + tan\ y}{1 - \tan x \tan y}.$$

Replacing y by $-y$ gives some further identities. These formulae are useful in conjunction with the standard triangles shown in Figure 1.5 for evaluating sines, cosines and tangents of angles $\pi/6$, $\pi/3$, $\pi/4$ and related angles.

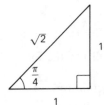 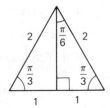

Fig. 1.5

EXERCISES 1.6.1

1 Sketch graphs of the sine, cosine and tangent functions. After what interval do the graphs repeat themselves?
2 Derive formulae for sin $2x$ and cos $2x$ in terms of sin x and cos x, and use the latter to derive formulae for sin $\frac{1}{2}x$ and cos $\frac{1}{2}x$ in terms of cos x.
3 Use the triangles in Figure 1.5 and the above identities to find the sine, cosine and tangent of

(i) $\pi/6$; (ii) $\pi/3$; (iii) $\pi/4$; (iv) $3\pi/4$; (v) $-5\pi/6$; (vi) $11\pi/3$.

1.7 EXPONENTIAL AND LOGARITHMIC FUNCTIONS

We deal with both of these functions in this section, since, as we shall see, they are closely connected. The exponential function can be defined in different ways; we give it here as the function exp, defined by

$$\exp(x) = 1 + x + x^2/2! + x^3/3! + \cdots + x^n/n! + \cdots.$$

This expression is called an infinite series: infinite because it contains an infinite number of terms. It can be proved that it converges—that is, it adds up to a finite sum—whatever value x takes. This means that we can take the domain of exp as $\mathbb{R}$. From a practical point of view, we obtain approximations by adding up the first few terms; the more we take, the more accurate is the answer. Try working out on your calculator the approximation $1 + 1 + 1/2! + 1/3! + \cdots + 1/10!$ to exp(1).

There is another representation of exp(x) which is useful because it reminds us of its properties. It is obtained by setting exp(1) = e (equal to 2.7183...); it is then possible to deduce (but you should not try it!) that exp(x) = e^x—that is, the number e raised to the power x, for any rational number x. We then define e^x to be exp(x) for any real number x. The important properties that

$$\exp(x + y) = \exp(x) \cdot \exp(y) \quad \text{and} \quad \exp(xy) = (\exp x)^y$$

11

become simply the index laws

$$e^{x+y} = e^x \cdot e^y \quad \text{and} \quad e^{xy} = (e^x)^y$$

Also, $\exp(0) = e^0 = 1$. From now on we shall use the form e^x rather than $\exp(x)$.

We now define the logarithmic or log function by

$$y = \log_e x \text{ if } y \text{ satisfies the equation } e^y = x$$

In words, this states that $\log_e x$ is the power to which e must be raised to obtain the value x. This is actually a definition of the logarithm to base e, the so-called natural logarithm. While deriving results, we shall use the notation $\log_e x$, which emphasises the base e, but thereafter we shall replace $\log_e x$ by the briefer notation $\ln x$.

For the definition to make sense, we see that x must be positive; in other words, we can only take the log of a positive number and so we must specify its domain as $\mathbb{R}^+$, which means the set of all positive real numbers. We also find that

$$y = \log_e x = \log_e(e^y) \quad \text{and} \quad x = e^y = e^{\log_e x}$$

Since $1 = e^0$, it follows that $\log_e 1 = 0$. The graphs of exp and $\log_e$ are sketched in Figure 1.6.

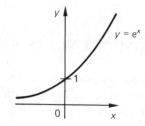
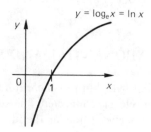

Fig. 1.6

From the relations above, we find that

$$e^{\log_e(xy)} = xy = e^{\log_e x} \cdot e^{\log_e y} = e^{\log_e x + \log_e y}$$

and so $\log_e(xy) = \log_e x + \log_e y$. If we put $y = x$ in this, we obtain $\log_e(x^2) = 2 \log_e x$; adding $\log_e x$ to itself n times, we see that $\log_e(x^n) = n \log_e x$ for any integer n. From the definition of $\log_e$ we obtain $x^n = e^{n \log_e x}$ for any positive integer n. We extend this to define x^t, where t is any real number, by $x^t = e^{t \log_e x}$ for any real number t. This enables us to attach a meaning to, and indeed to evaluate, quantities such as $3^{\sqrt{2}}$, which has no meaning in the normal sense of multiplying 3 by itself $\sqrt{2}$ times.

We now turn to more general exponential and logarithmic functions. *Let a be a fixed positive number.*

12

Then $y = a^x = e^{x \log_e a}$, so that for $a > 1$, $\log_e a > 0$, and the graph of a^x is similar in character to that of e^x. However, if $a < 1$, $\log_e a < 0$ and the graph of a^x now slopes in the other direction. In either case, the domain is $\mathbb{R}$, and the range is $\mathbb{R}^+$. We now define $\log_a$—that is, log to base a—by

$$y = \log_a x \text{ if } y \text{ satisfies the equation } a^y = x.$$

The domain and range of $\log_a$ are $\mathbb{R}^+$ and $\mathbb{R}$, respectively. We easily obtain the following relations, which are very similar to those for the exp and natural log functions:

$$a^{\log_a x} = x; \log_a a^y = y; \log_a xy = \log_a x + \log_a y; \log_a 1 = 0; \log_a x^y = y \log_a x$$

The graphs of a^x and $\log_a x$ for $a < 1$ are shown in Figure 1.7.

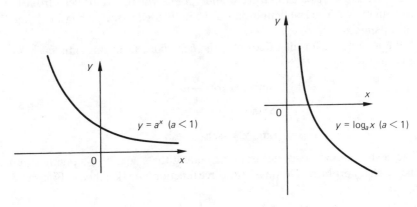

Fig. 1.7

Finally, because the definition of x^t involves $\log_e x$, we cannot evaluate it for $x = 0$. From the graph of the log function, we see that $\log_e x$ becomes large and negative as x approaches zero, so for $t > 0$, $e^{t \log_e x}$ becomes small; we therefore define $0^t = 0$ for $t > 0$. Note that 0^0, like $0/0$, is undefined.

The properties of $\log_e$ are obtained from those of $\log_a$ by putting $a = e$. The connection between them may be derived as follows: let $y = \log_a x$; then

$$x = a^y = e^{y \log_e a}$$

so that

$$\log_e x = y \log_e a = \log_a x \cdot \log_e a$$

giving

$$\log_a x = \log_e x / \log_e a$$

13

1.8 HYPERBOLIC FUNCTIONS

Before defining these, we note that if f and g are functions, then so is

$$f + g \text{ defined by } (f + g)(x) = f(x) + g(x)$$

Here we have defined the function $f + g$ as the function whose value at x is the sum of the values of the functions f and g at x. Similarly, we have the functions

$$f - g \quad \text{defined by} \quad (f - g)(x) = f(x) - g(x)$$

$$fg \quad \text{defined by} \quad (fg)(x) = f(x) \cdot g(x)$$

$$\frac{f}{g} \quad \text{defined by} \quad \left(\frac{f}{g}\right)(x) = \frac{f(x)}{g(x)}$$

We must as usual be careful to exclude points where $g(x)$ is zero from the domain of f/g. The last definition was actually used earlier, when we defined tan as sin/cos.

We now define for all values of x the *hyperbolic* functions sinh, cosh and tanh by

$$\sinh x = \tfrac{1}{2}(e^x - e^{-x});$$

$$\cosh x = \tfrac{1}{2}(e^x + e^{-x});$$

$$\tanh x = \sinh x / \cosh x$$

The cosh function arises naturally in finding the shape of a hanging chain, the catenary problem. Graphs of the three functions are sketched in Figure 1.8.

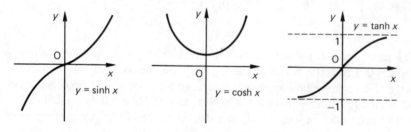

Fig. 1.8

It is also useful to know the reciprocal functions given by

$$\text{sech } x = 1/\cosh x$$

$$\text{cosech } x = 1/\sinh x \quad \text{for } x \neq 0$$

$$\coth x = 1/\tanh x \quad \text{for } x \neq 0$$

The following identities can easily be verified using the definitions:

$$\cosh^2 x - \sinh^2 x = 1$$

$$1 - \tanh^2 x = \text{sech}^2 x$$

$$\coth^2 x - 1 = \text{cosech}^2 x$$

$$\sinh(x + y) = \sinh x \cosh y + \cosh x \sinh y$$

$$\cosh(x + y) = \cosh x \cosh y + \sinh x \sinh y$$

$$\tanh(x + y) = (\tanh x + \tanh y)/(1 + \tanh x \tanh y)$$

A comparison with the trigonometric identities reveals a strong formal similarity. 'Osborne's rule' can be used to convert a trigonometric identity to a hyperbolic one: if a term contains *two* factors of sinh then change the sign. The term 'hyperbolic' arises from the parametric coordinates of points on a hyperbola. Recall that $\cos t$ and $\sin t$ represent the coordinates of a point on the unit circle, $x^2 + y^2 = 1$, corresponding to a sector of area $\frac{1}{2}t$. For hyperbolic functions, we consider instead the rectangular hyperbola whose equation is $x^2 - y^2 = 1$, as shown in Figure 1.9. It is clear that the point P, whose coordinates are $\cosh t$ and $\sinh t$, is on the hyperbola, since $\cosh^2 t - \sinh^2 t = 1$. It can be shown (see Chapter 3) that the shaded region has area $t/2$, thus establishing the analogue with a sector of a circle.

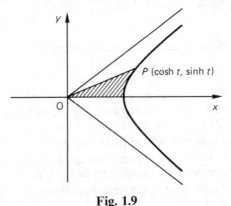

Fig. 1.9

EXERCISE 1.8.1

Sketch the graphs of (i) sech x; (ii) cosech x; (iii) coth x.

1.9 COMPOSITE AND INVERSE FUNCTIONS

Besides using arithmetic operations to form new functions, there are two other ways, which we discuss in this section.

The first of these is *composition* of functions. Suppose that we define a function g whose value at x is $y = g(x)$; now define a function f whose value at y is $f(y)$. Then the value of f at x is $f(g(x))$; we call $(f \circ g)$ the composition of g and f. We must not confuse this with the function we formed earlier as the product of f and g. The composite function $(f \circ g)$ is defined by $(f \circ g)(x) = f(g(x))$ and its domain is contained in that of g. For example, if $f(y) = \sin y$ and $y = g(x) = x^2$, then $(f \circ g)(x) = f(g(x)) = \sin x^2$. We note that the y appearing in the definition of f is just a dummy variable; we could equally well have defined f by $f(x) = \sin x$. The composite function $(g \circ f)$ is defined by $(g \circ f)(x) = g(f(x))$; for the example above, we have $(g \circ f)(x) = (\sin x)^2$. It is clear that, in general, $(f \circ g)$ is not the same function as $(g \circ f)$. (The phrase 'in general' is often used to qualify a statement in mathematics; it means that the statement is true except in a few special cases.)

The second new type of function is the inverse function. Suppose that if we can solve $y = f(x)$ for x, we find $x = g(y)$; then g is the *inverse function* of f, which we write as f^{-1}. *It is essential not to confuse this with the reciprocal of f,* whose value at x is

$$(f(x))^{-1} = \frac{1}{f(x)}$$

We look now at some examples to see what might happen when we try to find inverse functions.

Examples 1.9.1

1. Consider the function given by $f(x) = x^3$. We easily solve $y = x^3$ to find $x = y^{1/3}$. Thus, each value of y defines a unique value of x and the inverse function is given by $f^{-1}(y) = y^{1/3}$, or, in terms of x, $f^{-1}(x) = x^{1/3}$. This situation is clearly demonstrated by the graph of $y = x^3$ shown in Figure 1.10. To find the value of x corresponding to $y = y_1$, say, we find where the line $y = y_1$ intersects the graph of $y = x^3$. It is clear from the picture that there is only one point of intersection, whatever the value of y_1.
2. Now consider the function given by $g(x) = x(x - 2)(x + 2)$, whose graph is also shown in Figure 1.10. If we try to solve $y = x(x - 2)(x + 2)$ for $y = y_1$, with $-3 \le y_1 \le 3$, we find three possible values for x. In this case an inverse function does not exist, since we cannot define a unique value of x for every value of y.
3. Consider the function given by $f(x) = e^x$, whose graph is sketched in Figure 1.11. A line $y = y_1$ intersects this graph at a unique point if $y_1 > 0$, but not at all if $y \le 0$. We overcome this problem by taking the domain of f^{-1} to be $(0, \infty)$, which is just the *range of f*. We note that in this example solving $y = e^x$ for x gives $x = \log_e y$; so the inverse of the exponential function is the logarithm function. To sketch the graph of

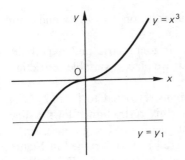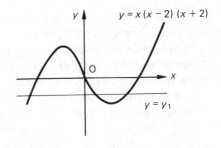

Fig. 1.10

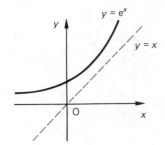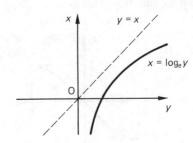

Fig. 1.11

f^{-1}, we must reverse the roles of x and y—that is, we put the x-axis in place of the y-axis, and vice versa; we obtain the graph by reflecting the graph of f in the line $y = x$, as shown in Figure 1.11.

We have seen that some functions do not have an inverse. Consider the function $y = x^2$, whose graph is shown in Figure 1.12. This cannot have an inverse, because for each non-zero value of y in the range of f, namely $y \geq 0$, there are two possible values of x. To overcome this, we *restrict* the function f to a domain such that, to each value of y, there is only one value of x. For example, we could restrict f to the domain $[0, \infty)$,

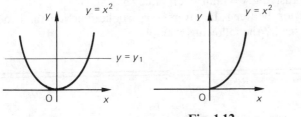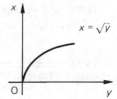

Fig. 1.12

in which case $y = x^2$ would have the unique solution $x = +\sqrt{y}$ and both f and f^{-1} have domain and range $[0, \infty)$.

We are now ready to define the inverse trigonometric and hyperbolic functions; all we have to do is make sure that we place a suitable restriction on the domain in order to define an inverse.

The graph of the sine function with domain restricted to $[-\pi/2, \pi/2]$ is sketched in Figure 1.13. It is clear that this restriction allows a line $y = y_1$ for $-1 \leq y_1 \leq 1$ to intersect the graph just once. The graph of $\sin^{-1} x$ with domain $[-1, 1]$ and range $[-\pi/2, \pi/2]$ is also sketched in Figure 1.13. Values of $\sin^{-1}$ in this range are called *principal values*. To evaluate $\sin^{-1} \frac{1}{2}$, for example, we require the angle which lies in $[-\pi/2, \pi/2]$ and whose sine is $\frac{1}{2}$. This is $\pi/6$.

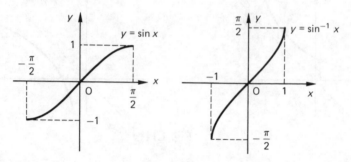

Fig. 1.13

Similarly, the domain of $\cos^{-1}$ is $[-1, 1]$ and its range of principal values is defined as $[0, \pi]$. For the tangent function, we take the domain to be the whole of $\mathbb{R}$ and define the range of principal values as $(-\pi/2, \pi/2)$, noting that the open interval is required here because the tangent is undefined at $\pm \pi/2$.

We now consider the inverse hyperbolic functions. As we saw in Section 1.8, the domain and range of sinh are both $(-\infty, \infty)$, so we define $\sinh^{-1} x$ as the value of y which satisfies $x = \sinh y$. Referring to Figure 1.8, we see that we must restrict the domain of cosh in order to invert it. To obtain the principal value, the restriction is $[0, \infty)$, so we define $\cosh^{-1} x$ as the non-negative value of y satisfying $x = \cosh y$. Since the range of cosh is $[1, \infty)$, we must take this as the domain of $\cosh^{-1}$. Finally, we define $\tanh^{-1} x$ as the value of y satisfying $x = \tanh y$ for values of x in $(-1, 1)$.

There is a connection between the inverse hyperbolic functions and the $\log_e$ function, as we see by the following example.

Example 1.9.2

Show that $\cosh^{-1} x = \ln(x + \sqrt{x^2 - 1})$ for $x > 1$.

18

Let $x = \cosh y = \frac{1}{2}(e^y + e^{-y})$. Then $2xe^y = (e^y)^2 + 1$, which is a quadratic in e^y. The solution is

$$e^y = x \pm \sqrt{x^2 - 1}$$

so that

$$y = \cosh^{-1} x = \ln(x \pm \sqrt{x^2 - 1})$$

Now

$$x - \sqrt{x^2 - 1} = \frac{1}{x + \sqrt{x^2 - 1}} < 1$$

since $x > 1$. We must therefore take the positive square root, since otherwise $\cosh^{-1} x$ would be the log of a number less than 1 and, hence, negative.

We have completed the description of the basic functions which may be used to construct our family of standard functions, which we now formally define. The family of functions obtained from the polynomial, trigonometric and exponential functions by applying arithmetic operations $(+, -, \cdot, /)$, composition and inversion of functions is called the *family of standard functions*.

Clearly, x^n is a member, since it is a polynomial, but is the function f given by $f(x) = x^a$, where a is a real number (such as $\sqrt{2}$), $x > 0$? It is not obvious that it is, since exponentiation is not included as an operation. However, we have $x^a = e^{a \ln x}$, and ln is a member, since it is the inverse of exp, so that x^a is indeed a member of our family. In particular, square, cube and nth roots of functions are in the family.

EXERCISES 1.9.1

1 By solving for x in terms of y, determine the inverse functions for the functions given by
   (i) $y = 3x - 2$; (ii) $y = e^{3x}$; (iii) $y = 1/x$, $x \neq 0$;
   (iv) $y = (x + 1)/(x - 2)$, $x \neq 2$.

2 With the range restricted to $[-\pi/2, \pi/2]$, which of the following statements are true?
   (i) $\sin^{-1} 0 = \pi$; (ii) $\sin^{-1}(1/\sqrt{2}) = \pi/4$;
   (iii) $\sin^{-1}(-\frac{1}{2}) = -\pi/6$; (iv) $\sin^{-1}(\sqrt{3}/2) = \pi/3$;
   (v) $\sin^{-1}\sqrt{3}$ is undefined.

3 Sketch the graphs of $\cos^{-1} x$ and $\tan^{-1} x$ and give their ranges of principal values.

4 Sketch the graphs of $\sinh^{-1} x$, $\cosh^{-1} x$ and $\tanh^{-1} x$ for suitable domains and give their ranges of principal values.

MISCELLANEOUS EXERCISES 1

1 Find the union and intersection of the sets

$$A = \{1, 2, 5, 8\}; \ B = \{2, 3, 6, 7, 8\}$$

If the universal set is $\{1, 2, 3, 4, 5, 6, 7, 8\}$, what are the sets $\setminus A$ and $\setminus B$?

2 Show that the function f defined by

$$f(x) = \begin{cases} x + 1, x \le 0 \\ x^2 + 1, x > 0 \end{cases}$$

is continuous at $x = 0$.

3 Find R and α so that

$$3 \cos \theta + 4 \sin \theta = R \sin(\theta + \alpha)$$

4 The displacement in m of a simple pendulum from the vertical is given approximately by $x = A \sin \omega t$, where t is the time in seconds, $\omega = \sqrt{g/l}$, g is gravitational acceleration in m/s^2 and l is the length of the pendulum in metres. Show that the displacement oscillates between $-A$ and A, and find its period (that is, the time taken for one complete oscillation).

5 The monthly mean temperature T at a resort has a maximum of $25\,°C$ in July and a minimum of $-5\,°C$ in January. Suppose that this variation is modelled by the formula $T = T_0 + a \cos(2\pi/p)(t - t_0)$, where t is the time in months, a is the amplitude of the variation in degrees C and p is the period in months. Find T_0, a and p. What will be the predicted mean temperature in September?

6 Find approximations to $e^{0.1}$ by calculating successively the sums of the first 1, 2, 3, ..., 10 terms of

$$e^x = 1 + x + x^2/2! + x^3/3! + \cdots$$

with $x = 0.1$. For each n, the $(n+1)$st term should be computed by multiplying the nth term by x/n, otherwise overflow is liable to occur.

7 A radioactive decay process is modelled by the formula $m = m_0 e^{-kt}$, where m is the mass at time t, m_0 is the mass at time zero and k is the decay constant. If m decays to $0.9m$ in 100 days, find the decay constant and hence how many days it takes for the mass to be halved (the *half-life*).

8 A chain suspended from two points of equal height takes the shape of a catenary $y = c\left(\cosh\left(\dfrac{x}{c}\right) - 1 \right)$, where x and y are the horizontal and vertical distances from the lowest point of the chain and c is a constant.

If the value of c is 100 m and the suspension points are 50 m apart, find the sag at the middle.

9   The heat flow through a cylindrical pipe is given by

$$q = K \frac{(\theta_1 - \theta_2)}{\ln(r_2/r_1)}$$

where θ_1, θ_2 are the temperatures inside and outside the pipe, r_1, r_2 are the radii of the inside and outside of the pipe and K is a constant. Solve this equation for r_1 in terms of the other quantities.

10   Explain why the function $f: \mathbb{R} \to \mathbb{R}$ defined by $f(x) = (x^2 - 1)/(x^2 + 1)$ does not have an inverse. Find f^{-1} when f is defined from $\mathbb{R}^+ \to [-1, 1)$.

1.10 ANSWERS TO EXERCISES

Exercise 1.1.1

(i) $[1, 3)$, (ii) $(-\infty, -5]$, (iii) $(-2, 1]$.

Exercises 1.2.1

1   (i) y not uniquely defined, (ii) is a function, (iii) y not uniquely defined, (iv) is a function, (v) is a function.
2   (i) $[0, \infty)$. (ii) $\mathbb{R} \backslash \{2\}$, (iii) $[-3, 3]$.
3   (i) $[-5, -1]$, (ii) $\mathbb{R}$, (iii) $[0, 1]$, (iv) $[0, \infty) = \mathbb{R}^+$, (v) $\mathbb{Z}$.

Exercises 1.6.1

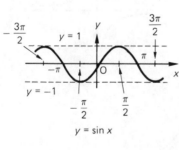

$y = \sin x$

$y = \cos x$

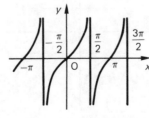

$y = \tan x$

21

1 Graphs repeat after a *period*, 2π for cos and sin, and π for tan.
2 $\sin 2x = \sin(x + x) = \sin x \cos x + \cos x \sin x = 2 \sin x \cos x$,
  $\cos 2x = \cos(x + x) = \cos x \cos x - \sin x \sin x = \cos^2 x - \sin^2 x$.
  $\cos 2x = \cos^2 x - \sin^2 x = 1 - 2 \sin^2 x = 2 \cos^2 x - 1$, so replacing x by
  $\dfrac{x}{2}$ and solving gives $\sin \dfrac{x}{2} = \sqrt{\dfrac{1}{2}(1 - \cos x)}$ and $\cos \dfrac{x}{2} = \sqrt{\dfrac{1}{2}(1 + \cos x)}$.

3

θ	$\dfrac{\pi}{6}$	$\dfrac{\pi}{3}$	$\dfrac{\pi}{4}$	$\dfrac{3\pi}{4}$	$-\dfrac{5\pi}{6}$	$\dfrac{11\pi}{3}$
$\sin \theta$	$\dfrac{1}{2}$	$\dfrac{\sqrt{3}}{2}$	$\dfrac{1}{\sqrt{2}}$	$\dfrac{1}{\sqrt{2}}$	$-\dfrac{1}{2}$	$-\dfrac{\sqrt{3}}{2}$
$\cos \theta$	$\dfrac{\sqrt{3}}{2}$	$\dfrac{1}{2}$	$\dfrac{1}{\sqrt{2}}$	$-\dfrac{1}{\sqrt{2}}$	$-\dfrac{\sqrt{3}}{2}$	$\dfrac{1}{2}$
$\tan \theta$	$\dfrac{1}{\sqrt{3}}$	$\sqrt{3}$	1	-1	$\dfrac{1}{\sqrt{3}}$	$-\sqrt{3}$

Exercise 1.8.1

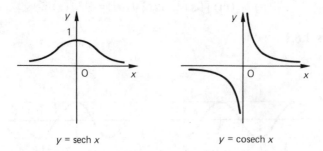

$y = \text{sech } x$ $y = \text{cosech } x$

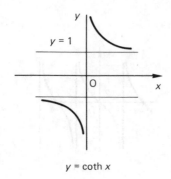

$y = \coth x$

Exercises 1.9.1

1   (i) $x = \dfrac{y+2}{3}$, (ii) $x = \dfrac{1}{3} \ln y$, (iii) $x = \dfrac{1}{y}$, $y \neq 0$, (iv) $x = \dfrac{2y+1}{y-1}$, $y \neq 1$.

2   (ii), (iii), (iv) and (v) are true.

3   Range of principal values $[0, \pi]$ for $\cos^{-1}$ and $\left(-\dfrac{\pi}{2}, \dfrac{\pi}{2} \right)$ for $\tan^{-1}$.

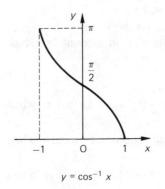

$y = \cos^{-1} x$

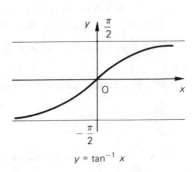

$y = \tan^{-1} x$

4   Range of principal values $\mathbb{R}$ for $\sinh^{-1}$ and $\tanh^{-1}$, $[0, \infty)$ for $\cosh^{-1}$.

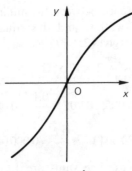

$y = \sinh^{-1} x$

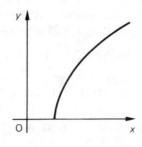

$y = \cosh^{-1} x$

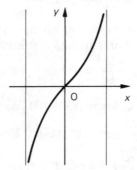

$y = \tanh^{-1} x$

23

Miscellaneous Exercises 1

1 $A \cup B = \{1, 2, 3, 5, 6, 7, 8\}, A \cap B = \{2, 8\}, \backslash A = \{3, 4, 6, 7\}, \backslash B = \{1, 4, 5\}.$

2 $\lim\limits_{x \to 0_+} f(x) = \lim\limits_{x \to 0_+} x^2 + 1 = 1 = f(0).$

3 $R \sin(\theta + \alpha) = R \sin \theta \cos \alpha + R \cos \theta \sin \alpha = 3 \cos \theta + 4 \sin \theta$, so $R \cos \alpha = 4$, $R \sin \alpha = 3$, giving $R^2 = R^2(\cos^2 \alpha + \sin^2 \alpha) = 4^2 + 3^2 = 5^2$, or $R = 5$. Also $\tan \alpha = \frac{3}{4}$ gives the value of α.

4 $\sin \omega t$ oscillates between ± 1, so x oscillates between $\pm A$. x goes through a complete oscillation as ωt increases by an amount 2π, which is as t increases by an amount $\dfrac{2\pi}{w}$.

5 Maximum temperature occurs when $\cos \dfrac{2\pi}{p}(t - t_0) = 1$, that is when $\dfrac{2\pi}{p}(t - t_0) = 0$, that is when $t = t_0$. The minimum occurs when $\cos \dfrac{2\pi}{p}(t - t_0) = -1$, which is when $\dfrac{2\pi}{p}(t - t_0) = \pi$, that is when $t - t_0 = \dfrac{p}{2}$. But $t - t_0 = 6$ (the number of months between July and January), so $p = 12$ months. The difference between the minimum and maximum temperatures is $25 - (-5) = 30\,°C$ and this equals $2a$, so $a = 15\,°C$. Putting $T = 25$ when $t - t_0 = 0$ gives $25 = T_0 + a$, so $T_0 = 10\,°C$. In September, $t = 2$, so $T = 10 + 15 \cos \dfrac{2\pi}{12} 2 = 17.5\,°C$.

6 The terms are $1,\ 0.1,\ 0.1 \times \dfrac{0.1}{2} = 0.005$, $0.005 \times \dfrac{0.1}{3} = 0.000\,166\,67$, $0.000\,166\,67 \times \dfrac{0.1}{4} = 0.000\,004\,17$, $0.000\,004\,17 \times \dfrac{0.1}{5} = 0.000\,000\,08$, the rest being all zero to 8 decimal places. The sums are 1, 1.1, 1.105, 1.105\,166\,67, 1.105\,170\,84, 1.105\,170\,92. The last is the value of $e^{0.1}$ correct to 8 decimal places.

7 $m = m_0$ when $t = 0$ and $m = 0.9 m_0$ when $t = 100$, so $0.9 m_0 = m_0 e^{-100k}$, so $-100k = \ln 0.9 = -0.1054$ giving $k = 0.001054$ days^{-1}. For the half-life, $\frac{1}{2} m_0 = m_0 e^{-kt}$, so $t = \dfrac{-\ln \frac{1}{2}}{k} \approx 658$ days $= 1$ year 293 days.

8 The equation of the catenary is $y = 100 \left\{ \cosh\left(\dfrac{x}{100}\right) - 1 \right\}$. At a suspension point, $x = 25$ m, so $y = 100\{\cosh 0.25 - 1\} = 3.141$ m.

9 $\ln\left(\dfrac{r_2}{r_1}\right) = (\theta_1 - \theta_2)\dfrac{K}{q} \Rightarrow \dfrac{r_2}{r_1} = e^{(\theta_1 - \theta_2)K/q} \Rightarrow r_1 = r_2 e^{-(\theta_1 - \theta_2)K/q}.$

10 Solving $y = \dfrac{x^2 - 1}{x^2 + 1}$ for x^2, we find $x^2 = \dfrac{1 + y}{1 - y}$. This is not defined for $y = 1$ and there are two possible answers when we take the square root for x. With f defined from $\mathbb{R}^+ \to [-1, 1)$ however $f^{-1}(y) = \sqrt{\dfrac{1 + y}{1 - y}}$, and f^{-1} is from $[-1, 1) \to \mathbb{R}^+$.

2 DIFFERENTIATION

2.1 INTRODUCTION

Consider an object moving in a straight line. Suppose that at time t it is a distance $s(t)$ from its position at time $t = 0$. What is its speed v at a given time t? If v is constant, then $v = \text{distance/time} = s(t)/t$. We see that $s(t) = vt$, with v constant, so that s is a function of time.

Constant speed motion is, however, very rare. Suppose that v varies with time, so that it can be considered as a function of time. The ratio $s(t)/t$ now becomes the average speed during the time t. We should like to be able to find the instantaneous value of the speed, $v(t)$, at time t. In order to obtain this, we consider the average speed over a small interval of time, δt, from t to $t + \delta t$. The distance travelled by the object during this interval is $\delta s = s(t + \delta t) - s(t)$, so that its average speed over the interval is

$$\frac{\delta s}{\delta t} = \frac{s(t + \delta t) - s(t)}{\delta t}$$

The smaller we make δt, the closer we should expect $\delta s / \delta t$ to become to the speed at t. In fact we really want δt to approach the value 0. If $\delta s / \delta t$ tends to a limit as $\delta t \to 0$, then the value of this limit is just the quantity we require, the speed at time t. We write

$$v(t) = \lim_{\delta t \to 0} \frac{\delta s}{\delta t} = \lim_{\delta t \to 0} \frac{s(t + \delta t) - s(t)}{\delta t}$$

Example 2.1.1

Compute the speed of a falling object, whose height at time t is $h(t) = 20 - \frac{1}{2}gt^2$.

At $t = 0$ the height of the object is $h(0) = 20$, so the distance travelled after time t is $s(t) = \frac{1}{2}gt^2$. Also $s(t + \delta t) = \frac{1}{2}g(t + \delta t)^2$, so $\delta s = s(t + \delta t) - s(t) = \frac{1}{2}g((t + \delta t)^2 - t^2) = g(t + \delta t/2)\delta t$. Thus,

$$\delta s/\delta t = g(t + \delta t/2)$$

and

$$v(t) = \lim_{\delta t \to 0} \frac{\delta s}{\delta t} = \lim_{\delta t \to 0} g(t + \delta t/2) = gt$$

We now consider the geometric problem of finding the slope of the tangent to the graph of a function f at a point x. We approximate the tangent at the point P, whose coordinates are $(x, f(x))$ by the chord joining P with the point Q whose coordinates are $(x + \delta x, f(x + \delta x))$ for a small δx. The slope of this chord, as we can see from Figure 2.1, is

$$\frac{f(x + \delta x) - f(x)}{\delta x}$$

We obtain the slope of the tangent by taking the limit of this slope as $\delta x \to 0$.

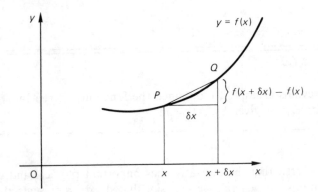

Fig. 2.1

We write this as

$$\lim_{\delta x \to 0} \frac{f(x + \delta x) - f(x)}{\delta x}$$

Definition

Suppose that a function f is defined on an interval containing x and let $y = f(x)$. If the ratio

$$\frac{\delta y}{\delta x} = \frac{f(x + \delta x) - f(x)}{\delta x}$$

27

tends to a finite limit as δx tends to zero, then we say that f is *differentiable* at x with *derivative* equal to that limit. We write

$$\frac{dy}{dx} = f'(x) = \lim_{\delta x \to 0} \frac{f(x + \delta x) - f(x)}{\delta x}$$

It is useful to introduce the idea of a *differential operator*, d/dx. This simply performs the operation of differentiation. Thus, we shall write

$$\frac{d}{dx} f(x) = f'(x)$$

or sometimes just

$$\frac{d}{dx} f = f'$$

Note that an arbitrary general function need not have a derivative at a given point, but functions in our standard family will turn out to be differentiable at almost every point of their domains. As the notation implies, the derivative f' of f is also a function; it is defined by its value, $f'(x)$, at each value of x for which it exists.

Example 2.1.2

In our previous example we obtained the function s, given by $s(t) = \frac{1}{2}gt^2$. Its derivative, s', is given by $s'(t) = gt$.

We have seen that the derivative has important physical and geometric interpretations. In general, we can say that dy/dx represents the rate of change of y with respect to x at a certain value of x.

For many functions it can be quite hard to evaluate dy/dx using the definition in terms of a limit (we call this 'differentiating from first principles'). Fortunately, for functions in our standard family, there is a way to avoid working out these limits. Recall that every function in our standard family can be obtained from a few relatively simple functions by function-forming operations. Thus, if we know the derivatives of x, $\sin x$, e^x and the constant functions and if we have rules telling us how derivatives behave with respect to the four arithmetic operations, composition of functions and inverse function forming, then we can work out the derivative of any function in our family. This we shall now start to do.

If $y = c$, a constant, then its value is the same for every value of x, so that $\delta y = 0$ and therefore $dy/dx = 0$ for all values of x.

For the function f, given by $f(x) = x$, we have $y = f(x)$ and $\delta y = f(x + \delta x) - f(x) = x + \delta x - x = \delta x$, giving $\delta y / \delta x = 1$ and hence $dy/dx = 1$ for all values of x.

Before considering derivatives of the sine and exponential functions, let us establish the rules for differentiating various combinations of functions.

EXERCISE 2.1.1

Find the derivatives of $x^2 - x + 1$ and x^3 from first principles.

2.2 DERIVATIVES OF COMBINATIONS OF FUNCTIONS

Let u and v be differentiable functions of x and consider the function $u + v$. The change in $u + v$ caused by a change in x of δx is given by

$$\delta(u + v) = (u(x + \delta x) + v(x + \delta x)) - (u(x) + v(x))$$

$$= (u(x + \delta x) - u(x)) + (v(x + \delta x) - v(x))$$

$$= \delta u + \delta v$$

where δu and δv are the changes in u and v, respectively, caused by a change δx in x. Thus,

$$\frac{\delta(u + v)}{\delta x} = \frac{\delta u}{\delta x} + \frac{\delta v}{\delta x}$$

and when we take the limit, this becomes

$$\frac{d}{dx}(u + v) = \frac{du}{dx} + \frac{dv}{dx}$$

We can express this rule more neatly as

$$(u + v)' = u' + v'$$

and add two other rules, which may be derived in a similar way,

$$(u - v)' = u' - v'$$

$$(cu)' = cu'$$

where c is a constant.

Let us now look at the product of two functions, u and v. Put $y = u(x)v(x)$. Then

$$\delta y = u(x + \delta x)v(x + \delta x) - u(x)v(x)$$

$$= u(x + \delta x)(v(x + \delta x) - v(x)) + v(x)(u(x + \delta x) - u(x))$$

29

giving

$$\frac{\delta y}{\delta x} = u(x + \delta x)\frac{\delta v}{\delta x} + v(x)\frac{\delta u}{\delta x}$$

Now, letting $\delta x \to 0$, and assuming that the differentiable function u is automatically continuous, so that $u(x + \delta x) \to u(x)$ as $\delta x \to 0$, we obtain the *product rule*

$$\frac{d}{dx}u(x)v(x) = u(x)\frac{dv}{dx} + v(x)\frac{du}{dx}$$

Now let

$$y = \frac{u(x)}{v(x)}$$

Then

$$\delta y = \frac{u(x + \delta x)}{v(x + \delta x)} - \frac{u(x)}{v(x)}$$

from which we obtain by taking over a common denominator and rearranging

$$\delta y = \frac{v(x)(u(x + \delta x) - u(x)) - u(x)(v(x + \delta x) - v(x))}{v(x)v(x + \delta x)}$$

Introducing δu and δv and dividing by δx, we obtain

$$\frac{\delta y}{\delta x} = \frac{v(x)\dfrac{\delta u}{\delta x} - u(x)\dfrac{\delta v}{\delta x}}{v(x)v(x + \delta x)}$$

Letting $\delta x \to 0$, we obtain the *quotient rule*,

$$\frac{d}{dx}\left\{\frac{u(x)}{v(x)}\right\} = \frac{v(x)\dfrac{du}{dx} - u(x)\dfrac{dv}{dx}}{v(x)^2}$$

We state these rules as:

If u, v are differentiable functions of x, then so are uv and u/v and their derivatives are given by

$$(uv)' = uv' + u'v$$

and

$$\left(\frac{u}{v}\right)' = \frac{u'v - uv'}{v^2}$$

A special case of the last rule is worth setting out separately. If $u = 1$, then

$u' = 0$ and so

$$\left(\frac{1}{v}\right)' = -\frac{v'}{v^2}$$

Example 2.2.1

Use the product rule to obtain the derivatives of x^2, x^3, ..., x^n, for integer n.
For the derivative of x^2, we take $u = v = x$ in the product rule to obtain

$$\frac{d}{dx}(x \cdot x) = x \cdot 1 + 1 \cdot x = 2x$$

For x^3, we take $u = x$ and $v = x^2$ to obtain

$$\frac{d}{dx}(x \cdot x^2) = x \cdot 2x + 1 \cdot x^2 = 3x^2$$

Finding the derivative of x^n requires use of the technique of proof known as 'mathematical induction'; to prove that a statement is true for all positive integers n, we must prove that it is true for $n = 1$, then that, if it is true for $n = k$, it is true for $n = k + 1$. From the derivatives we have found for x^2 and x^3, we suspect that the derivative of x^n is nx^{n-1}. We already know this is true for $n = 1$, so let us assume that it is true for $n = k$—that is,

$$\frac{d}{dx}x^k = kx^{k-1}$$

Now take $u = x$ and $v = x^k$ in the product rule to obtain

$$\frac{d}{dx}x^{k+1} = \frac{d}{dx}(x \cdot x^k)$$

$$= x \cdot kx^{k-1} + 1 \cdot x^k$$

$$= (k+1)x^k$$

This is one of the most important rules of differentiation; it is valid not only for all positive integers, but also for *all* real numbers, as we shall see in Section 2.4.

Let f and g be two different functions and suppose that we wish to find the derivative of the composite function $f \circ g$. Put $y = g(x)$ and $z = f(y)$, so that $z = f(g(x))$ (z is a 'function of a function' of x). Then

$$\frac{\delta z}{\delta x} = \frac{f(y + \delta y) - f(y)}{\delta x}$$

$$= \frac{f(y + \delta y) - f(y)}{\delta y} \cdot \frac{\delta y}{\delta x}$$

$$= \frac{\delta z}{\delta y} \cdot \frac{\delta y}{\delta x}$$

As $\delta x \to 0$, then $\delta y = g(x + \delta x) - g(x) \to 0$ also, by the continuity of g, so in the limit we obtain the *chain rule*,

$$\frac{dz}{dx} = \frac{dz}{dy} \cdot \frac{dy}{dx}$$

The argument above is not valid if $\delta y = 0$, since then we could not divide by it. However, in this case y would be constant and, hence, so would $z = f(y)$. The chain rule would thus give the correct answer of 0 in this case.

We can also express the chain rule in the alternative form

$$(f(g(x))' = f'(g(x)) \cdot g'(x)$$

Example 2.2.2

Differentiate $z = (x^2 + 3)^7$.

Let $y = x^2 + 3$, so that $dy/dx = 2x$. Expressing z in terms of y, we have $z = y^7$, giving $dz/dy = 7y^6$. The chain rule gives

$$\frac{dz}{dx} = \frac{dz}{dy} \cdot \frac{dy}{dx}$$

$$= 7y^6 \cdot 2x$$

Substituting for y in terms of x and simplifying, we find

$$\frac{dz}{dx} = 14x(x^2 + 3)^6$$

Now suppose that f and g are functions inverse to each other. Putting $y = g(x)$, we have $x = g^{-1}(y) = f(y) = f(g(x))$. Applying the chain rule to this, we find that

$$1 = \frac{dx}{dx} = (f(g(x))' = f'(g(x)) \cdot g'(x) = f'(y) \cdot g'(x)$$

so that

$$\frac{dx}{dy} = f'(y) = \frac{1}{g'(x)} = \frac{1}{\dfrac{dy}{dx}}$$

This is called the *inverse function rule*. It is best remembered in the form of the identity

$$\frac{dx}{dy} = \frac{1}{\dfrac{dy}{dx}}$$

We illustrate it with an example.

Example 2.2.3

Differentiate $\sqrt{x}$.

Write $y = \sqrt{x}$, so that $x = y^2$. Therefore, $dx/dy = 2y$, so the rule gives

$$\frac{dy}{dx} = \frac{1}{2y} = \frac{1}{2\sqrt{x}}$$

EXERCISES 2.2.1

1 Differentiate

(i) $(x + 1)(x^2 - x + 1)$; (ii) $\dfrac{5}{x^2 - 1}$; (iii) $\dfrac{x^2 + x + 1}{x^2 - x + 1}$.

2 Use the inverse function rule to differentiate $x^{1/3}$.

3 Differentiate

(i) $\sqrt{x^2 + 1}$; (ii) $\dfrac{\sqrt{x}}{1 + x^2}$; (iii) $\sqrt{\dfrac{1 + x}{1 - x}}$.

2.3 DERIVATIVES OF TRIGONOMETRIC FUNCTIONS

Our next task is to find the derivative of the sine function. As a preliminary, we shall need to find

$$\lim_{x \to 0} \frac{\sin x}{x}$$

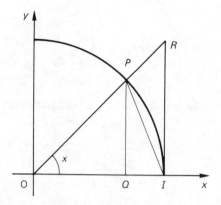

Fig. 2.2

Here both the numerator and denominator tend to zero, so care must be taken in evaluating the ratio. We shall use the geometric definition of trigonometric functions in a sector of the unit circle, as in Figure 2.2. Here $OI = OP = 1$ and P has coordinates $(\cos x, \sin x)$. The areas of triangles OPI and ORI are $\frac{1}{2}\sin x$, $\frac{1}{2}\tan x$, respectively, and the area of the sector OPI of the unit circle is $\frac{1}{2}x$. A comparison of these areas gives

$$\frac{\sin x}{2} < \frac{x}{2} < \frac{\tan x}{2}$$

which, when multiplied through by $2/\sin x$ becomes

$$1 < \frac{x}{\sin x} < \frac{1}{\cos x}$$

Taking reciprocals, we obtain

$$1 > \frac{\sin x}{x} > \cos x$$

which becomes in the limit as $x \to 0$

$$1 \geq \lim_{x \to 0} \frac{\sin x}{x} \geq 1$$

Hence,

$$\lim_{x \to 0} \frac{\sin x}{x} = 1$$

Now using the fourth trigonometric identity in Section 1.6 with x replaced by $x + \delta x$ and y replaced by $-x$, we have

$$\frac{\sin(x + \delta x) - \sin x}{\delta x} = \frac{\sin \frac{1}{2}\delta x}{\frac{1}{2}\delta x} \cdot \cos(x + \frac{1}{2}\delta x)$$

In the limit as $\delta x \to 0$, the left-hand member becomes $d/dx(\sin x)$, while, using the above limit with x replaced by $\frac{1}{2}\delta x$ in the right-hand member, together with $\cos(x + \frac{1}{2}\delta x) \to \cos x$ as $\delta x \to 0$ (from the continuity of the cosine function), we obtain

$$\frac{d}{dx} \sin x = \cos x$$

From this, the derivatives of all other trigonometric functions and their inverses can be deduced, using the rules of differentiation. For example, to find the derivative of the cosine function, set $y = \cos x$ and write this as $y = \sin z$, where $z = \pi/2 - x$. The chain rule now gives

$$\frac{dy}{dx} = \frac{dy}{dz} \cdot \frac{dz}{dx} = \cos z \cdot (-1) = -\cos\left(\frac{\pi}{2} - x\right) = -\sin x$$

Thus,

$$\frac{d}{dx} \cos x = -\sin x$$

Next let $y = \sin^{-1} x$. Then $x = \sin y$, and using the inverse function rule,

$$\frac{dy}{dx} = \frac{1}{\dfrac{dx}{dy}} = \frac{1}{\cos y} = \frac{1}{\sqrt{1 - \sin^2 y}} = \frac{1}{\sqrt{1 - x^2}}$$

We have taken the positive square root because $y = \sin^{-1} x$ is normally assumed to have a principal value between $-\pi/2$ and $+\pi/2$, and then $\cos y > 0$. Thus,

$$\frac{d}{dx} \sin^{-1} x = \frac{1}{\sqrt{1 - x^2}}$$

EXERCISES 2.3.1

1 Find the derivatives of

(i) $\cos 3x$; (ii) $\sin^2 x$; (iii) $\sin(x^2)$; (iv) $\tan x$.

2 Show that $\dfrac{d}{dx} \tan^{-1} x = \dfrac{1}{1 + x^2}$.

2.4 DERIVATIVES OF EXPONENTIAL AND LOGARITHMIC FUNCTIONS

We now turn to the exponential and logarithmic functions. Recall that the exponential function, exp, is defined by

$$\exp(x) = e^x = 1 + x + \frac{x^2}{2!} + \frac{x^3}{3!} + \cdots + \frac{x^n}{n!} + \cdots$$

Now $(d/dx)(x^n/n!) = (x^{n-1})/(n-1)!$; in other words, each term when differentiated becomes equal to the previous term, and the whole series is reproduced. We expect, therefore, that

$$\frac{d}{dx} e^x = e^x$$

This result can be proved formally, but it is beyond our scope. Taking now $y = \log_e x$, we have $x = e^y$, so that $dx/dy = e^y = x$. Hence $dy/dx = 1/x$. So

$$\frac{d}{dx} \log_e x = \frac{1}{x}$$

Let a be a real number. Then, for $x > 0$, we have, with the help of the chain rule,

$$\frac{d}{dx} x^a = \frac{d}{dx} e^{a\log_e x} = \frac{a}{x} e^{a\log_e x} = ax^{a-1}$$

Thus,

$$\frac{d}{dx} x^a = ax^{a-1}$$

for any a, which generalises the earlier rule for differentiating integer powers of x.

We can use a similar technique to differentiate functions of the form $f(x)^{g(x)}$. We write such a function as $e^{g(x)\cdot\log_e f(x)}$ and use the chain rule to carry out the differentiation, as the following example illustrates.

Example 2.4.1

Differentiate $y = x^x$.

We write $y = e^{x\log_e x} = e^z$, say, where $z = x \log_e x$. Then

$$\frac{dy}{dx} = \frac{dy}{dz}\frac{dz}{dx} = e^z \left(\log_e x + x \cdot \frac{1}{x} \right)$$

$$= (\log_e x + 1)x^x$$

An alternative method of differentiating $f(x)^{g(x)}$ is to take logs and then use the technique of *implicit differentiation*. We obtain $\log_e y = g(x)\log_e f(x)$, which we wish to differentiate with respect to x. Writing $p = \log_e y$, we use the chain rule

$$\frac{dp}{dx} = \frac{dp}{dy}\frac{dy}{dx} = \frac{1}{y}\frac{dy}{dx}$$

$$\frac{d}{dx}\log_e y = \frac{1}{y}\frac{dy}{dx}$$

We differentiate the right-hand side, $g(x)\log_e f(x)$, in the usual way.

Example 2.4.2

We use this technique to differentiate $y = x^x$ again.

Taking logs gives us $\log_e y = x \log_e x$, and, differentiating with respect to x, we obtain

$$\frac{1}{y}\frac{dy}{dx} = \log_e x + x\,\frac{1}{x}$$

giving

$$\frac{dy}{dx} = (\log_e x + 1)y = (\log_e x + 1)x^x$$

You can elect to use either method of solving this example.

A list of derivatives of standard functions and some other useful ones is given at the end of the chapter.

EXERCISES 2.4.1

1 Use the definition of a^x to show that for $a > 0$ $d/dx(a^x) = a^x \log_e a$.

2 Use the definitions of the hyperbolic functions to show that

(i) $\dfrac{d}{dx}\sinh x = \cosh x$; (ii) $\dfrac{d}{dx}\cosh x = \sinh x$; (iii) $\dfrac{d}{dx}\tanh x = \operatorname{sech}^2 x$.

3 Use the inverse function rule to show that for $a > 0$

$$\frac{d}{dx}\sinh^{-1}\frac{x}{a} = \frac{1}{\sqrt{a^2 + x^2}}.$$

4 Show that

$$\frac{d}{dx} \log_a x = \frac{1}{x \log_e a}$$

for $a, x > 0$.

5 Differentiate

(i) $3^{\sin x}$; (ii) $\cos(3^x)$.

2.5 HIGHER DERIVATIVES

Suppose $y = \sin 2x$. Then $dy/dx = 2 \cos 2x$ gives a new function of x, and we can differentiate it again, to obtain

$$\frac{d}{dx}\left(\frac{dy}{dx}\right) = -4 \sin 2x$$

We abbreviate the left-hand side to d^2y/dx^2, or if $y = f(x)$, to $f''(x)$. In general, given $y = f(x)$, we can calculate successive derivatives

$$\frac{dy}{dx}, \frac{d^2y}{dx^2}, \ldots, \frac{d^n y}{dx^n}, \ldots, \text{ or } f'(x), f''(x) = f^{(2)}(x), \ldots, f^{(n)}(x), \ldots$$

(It is sometimes convenient to extend this notation to use $f^{(0)}$ and $f^{(1)}$ to represent f and f', respectively.)

Examples 2.5.1

1. Let $f(x) = x^3 + x^2 - 3$. Then

$$f'(x) = 3x^2 + 2x, \ f''(x) = 6x + 2, \ f^{(3)}(x) = 6, \text{ etc.}$$

2. Find $f^{(4)}(x)$ when $f(x) = \sin x$.

$$f'(x) = \cos x, \ f''(x) = -\sin x, \ f^{(3)}(x) = -\cos x, \ f^{(4)}(x) = \sin x.$$

We now consider the successive derivatives of a product function $y = uv$. By the product rule

$$y' = (uv)' = u'v + uv'$$

We differentiate again, using the product rule on $u'v$ and uv', to obtain

$$y'' = (u''v + u'v') + (u'v' + uv'') = u''v + 2u'v' + uv''$$

Differentiating again gives

$$y^{(3)} = u^{(3)}v + 3u''v' + 3u'v'' + uv^{(3)}$$

We note that the coefficients in the expressions for y'' and $y^{(3)}$ are exactly those occurring in the binomial expansions

$$(a+b)^2 = a^2 + 2ab + b^2$$

and

$$(a+b)^3 = a^3 + 3a^2b + 3ab^2 + b^3$$

The nth derivative also conforms to this pattern and is given by *Leibniz' Rule*:

$$(uv)^{(n)} = u^{(n)}v + \binom{n}{1}u^{(n-1)}v^{(1)} + \cdots + \binom{n}{r}u^{(n-r)}v^{(r)} + \cdots + uv^{(n)}$$

where the binomial coefficients are given by

$$\binom{n}{r} = \frac{n!}{(n-r)!r!}$$

Example 2.5.2

Find

$$\frac{d^4}{dx^4}(e^{2x}\sin x).$$

It is convenient to write down the binomial coefficients, the derivatives of $u = e^{2x}$ (in reverse order) and the derivatives of $v = \sin x$ in columns:

Binomial coefficients	Derivatives of u	Derivatives of v
1	$2^4 e^{2x}$	$\sin x$
4	$2^3 e^{2x}$	$\cos x$
6	$2^2 e^{2x}$	$-\sin x$
4	$2 e^{2x}$	$-\cos x$
1	e^{2x}	$\sin x$

Thus,

$$\frac{d^4}{dx^4}(e^{2x}\sin x) = 16e^{2x}\sin x + 32e^{2x}\cos x - 24e^{2x}\sin x - 8e^{2x}\cos x + e^{2x}\sin x$$

$$= -7e^{2x}\sin x + 24e^{2x}\cos x$$

$$= e^{2x}(24\cos x - 7\sin x)$$

EXERCISES 2.5.1

1 Find the first three derivatives of

(i) $x^3(x^2 - 1)$; (ii) $1/x$; (iii) $x \sin x$; (iv) $\cosh x$.

2 Find

(i) $\dfrac{\mathrm{d}^5}{\mathrm{d}x^5}(e^x \cos 3x)$; (ii) $\dfrac{\mathrm{d}^{10}}{\mathrm{d}x^{10}}(x^3 e^x)$.

2.6 POWER SERIES EXPANSIONS

A *power series* is a series whose terms consist of multiples of powers of x; its general form is

$$a_0 + a_1 x + a_2 x^2 + \cdots$$

where the *coefficients*, $a_0, a_1, a_2, \ldots$, are numbers. If it converges, its sum will give a function of x. Often we wish, not to find the sum of the series, but to find a power series whose sum represents a given function, f. If we can find such a series, we call it the *power series expansion* of f. There are two problems in finding power series expansions: (1) finding the coefficients, (2) proving convergence. The latter is beyond our scope here, but we give a useful result.

Every power series in x has a *radius of convergence*, R, which is either a non-negative number or $+\infty$; the series converges if $-R < x < R$ and diverges if $x < -R$ or $x > R$. The general theory does not say what happens if $x = R$ or $-R$. The interval $(-R, R)$ is called the *interval of convergence*, and, for values of x in this interval, we may differentiate or integrate the series 'term by term'.

We can find the coefficients of a power series expansion by equating its derivatives at $x = 0$ to corresponding derivatives of the function, also at $x = 0$. Suppose that

$$f(x) = a_0 + a_1 x + \cdots + a_n x^n + a_{n+1} x^{n+2} + \cdots$$

Then, differentiating n times,

$$f^{(n)}(x) = 0 + 0 + \cdots + a_n n(n-1)(n-2) \ldots 2 \cdot 1$$
$$+ a_{n+1}(n+1)n(n-1) \ldots 3.2x + \cdots$$

Putting $x = 0$, leaves only the term $a_n n!$, and we obtain

$$f^{(n)}(0) = a_n n!$$

so that

$$a_n = \frac{f^{(n)}(0)}{n!}$$

These coefficients can be found if we can differentiate f as often as we like, but this does not guarantee convergence. We list the results of using this formula on a number of functions in the table below. Also included are the intervals of convergence.

	Interval of convergence
$e^x = 1 + x + \dfrac{x^2}{2!} + \cdots + \dfrac{x^n}{n!} + \cdots$	$(-\infty, \infty)$
$\cos x = 1 - \dfrac{x^2}{2!} + \dfrac{x^4}{4!} - \cdots + (-1)^n \dfrac{x^{2n}}{(2n)!} + \cdots$	$(-\infty, \infty)$
$\sin x = x - \dfrac{x^3}{3!} + \dfrac{x^5}{5!} - \cdots + (-1)^n \dfrac{x^{2n+1}}{(2n+1)!} + \cdots$	$(-\infty, \infty)$
$\cosh x = 1 + \dfrac{x^2}{2!} + \dfrac{x^4}{4!} + \cdots + \dfrac{x^{2n}}{(2n)!} + \cdots$	$(-\infty, \infty)$
$\sinh x = x + \dfrac{x^3}{3!} + \dfrac{x^5}{5!} + \cdots + \dfrac{x^{2n+1}}{(2n+1)!} + \cdots$	$(-\infty, \infty)$
$\ln(1+x) = x - \dfrac{x^2}{2} + \dfrac{x^3}{3} - \cdots + (-1)^{n+1} \dfrac{x^n}{n} + \cdots$	$(-1, 1)$
$(1+x)^k = 1 + kx + k(k-1)\dfrac{x^2}{2!} + \cdots$	
$+ \dfrac{k(k-1)(k-2) \ldots (k-n+1)}{n!} x^n + \cdots$	$(-1, 1)$
$\tan^{-1} x = x - \dfrac{x^3}{3} + \dfrac{x^5}{5} + \cdots + (-1)^n \dfrac{x^{2n+1}}{2n+1} + \cdots$	$(-1, 1)$

EXERCISES 2.6.1

1 Derive the above power series by differentiating repeatedly and putting $x = 0$ to find the coefficients.
2 Differentiate term by term the first five power series above, and notice what function the resulting series represents.

2.7 LIST OF STANDARD DERIVATIVES

In the following table, x and a are real numbers, n is an integer and the last column gives any restrictions which apply (k is an integer).

$$\frac{d}{dx} a = 0 \qquad\qquad a \text{ constant}$$

$$\frac{d}{dx} x = 1$$

$$\frac{d}{dx} x^n = nx^{n=1} \qquad\qquad n \neq 0; \text{ if } n < 0, \text{ then } x \neq 0$$

$$\frac{d}{dx} e^x = c^x$$

$$\frac{d}{dx} \ln x = \frac{1}{x} \qquad\qquad x > 0$$

$$\frac{d}{dx} a^x = a^x \ln a \qquad\qquad a > 0$$

$$\frac{d}{dx} \log_a x = \frac{1}{x \ln a} \qquad\qquad a, x > 0$$

$$\frac{d}{dx} x^a = ax^{a-1} \qquad\qquad x > 0$$

$$\frac{d}{dx} \sin x = \cos x$$

$$\frac{d}{dx} \cos x = -\sin x$$

$$\frac{d}{dx} \tan x = \sec^2 x \qquad\qquad x \neq k\pi + \frac{\pi}{2}$$

$$\frac{d}{dx} \cot x = -\mathrm{cosec}^2 x \qquad\qquad x \neq k\pi$$

$$\frac{d}{dx} \sec x = \sec x \tan x \qquad\qquad x \neq k\pi + \frac{\pi}{2}$$

$$\frac{d}{dx} \mathrm{cosec}\, x = -\mathrm{cosec}\, x \cot x \qquad\qquad x \neq k\pi$$

$$\frac{d}{dx} \sin^{-1} x = \frac{1}{\sqrt{1 - x^2}} \qquad\qquad -1 < x < 1$$

$$\frac{d}{dx}\cos^{-1}x = -\frac{1}{\sqrt{1-x^2}} \qquad -1 < x < 1$$

$$\frac{d}{dx}\tan^{-1}x = \frac{1}{1+x^2}$$

$$\frac{d}{dx}\cot^{-1}x = -\frac{1}{1+x^2}$$

$$\frac{d}{dx}\sec^{-1}x = \frac{1}{|x|\sqrt{x^2-1}} \qquad |x| > 1$$

$$\frac{d}{dx}\operatorname{cosec}^{-1}x = -\frac{1}{|x|\sqrt{x^2-1}} \qquad |x| > 1$$

$$\frac{d}{dx}\sinh x = \cosh x$$

$$\frac{d}{dx}\cosh x = \sinh x$$

$$\frac{d}{dx}\tanh x = \operatorname{sech}^2 x$$

$$\frac{d}{dx}\coth x = -\operatorname{cosech}^2 x \qquad x \neq 0$$

$$\frac{d}{dx}\operatorname{cosech} x = -\operatorname{cosech} x \coth x \qquad x \neq 0$$

$$\frac{d}{dx}\operatorname{sech} x = -\operatorname{sech} x \tanh x$$

$$\frac{d}{dx}\sinh^{-1}x = \frac{1}{\sqrt{x^2+1}}$$

$$\frac{d}{dx}\cosh^{-1}x = \frac{1}{\sqrt{x^2-1}} \qquad x > 1$$

$$\frac{d}{dx}\tanh^{-1}x = \frac{1}{1-x^2} \qquad x < 1$$

$$\frac{d}{dx}\coth^{-1}x = \frac{1}{1-x^2} \qquad x > 1$$

$$\frac{d}{dx}\operatorname{sech}^{-1}x = -\frac{1}{x\sqrt{1-x^2}} \qquad 0 < x < 1$$

$$\frac{d}{dx}\operatorname{cosech}^{-1}x = -\frac{1}{x\sqrt{1+x^2}} \qquad x \neq 0$$

MISCELLANEOUS EXERCISES 2

1 The length l of a rod in metres at temperature T in degrees C is given by $l = 1 + 0.000\,012\,T + 0.000\,000\,11\,T^2$. Find the rate at which l increases with T when $T = 100\,°C$.

2 The distance s in metres of a body from a fixed point at time t in seconds is given by $s = 20t - 5t^2$. What is its velocity at time $100\,s$?

3 The period T in seconds of a simple pendulum of length l m is given by $T = 2\pi\sqrt{l/g}$, where g m/s^2 is the acceleration due to gravity. Find the rate of change of period with length when $l = 1$ m.

4 Water is poured into a right circular cone of semi-angle 45°, with its axis vertical, at a rate of 10cc per second. At what speed is the surface of the water rising when the depth of the water is 10 cm? (Hint: when the depth of water is x cm, the volume Vcc of water is given by $V = (\pi/3)x^3$; differentiate this with respect to time.)

5 A ship leaves port sailing due north at 12 knots. Another ship leaves the same port an hour later, sailing due east at 16 knots. How fast are they separating four hours after the first ship left port?

6 The distances u and v of an object and its image from a lens of focal length f are connected by the formula $1/u + 1/v = 1/f$. An object is moved towards a lens of focal length 4 cm at a speed of 6 cm per second. Find how fast the image recedes from the lens when the object is 5 cm from the lens.

7 The field strength H of a magnet of length $2l$ and moment M at a point on its axis distance x from its centre is given by

$$H = \frac{M}{2l}\left[\frac{1}{(x-l)^2} - \frac{1}{(x+l)^2}\right]$$

Assuming that l is small compared to x, expand each of the terms in brackets as a power series and show that H is approximately $2M/x^3$.

2.8 ANSWERS TO EXERCISES

Exercise 2.1.1

Let $f(x) = x^2 - x + 1$, then

$$f(x + \delta x) - f(x) = (x + \delta x)^2 - (x + \delta x) + 1 - (x^2 - x + 1) = 2x\delta x + \delta x^2 - \delta x,$$

so

$$\frac{f(x + \delta x) - f(x)}{\delta x} = 2x + \delta x - 1 \to 2x - 1 \quad \text{as} \quad \delta x \to 0.$$

Let $f(x) = x^3$, then

$$\frac{f(x + \delta x) - f(x)}{\delta x} = \frac{(x + \delta x)^3 - x^3}{\delta x} = 3x^2 + 3x\delta x + \delta x^2 \to 3x^2 \quad \text{as} \quad \delta x \to 0.$$

Exercises 2.2.1

1 (i) $3x^2$, (ii) $\dfrac{-10x}{(x^2 - 1)^2}$, (iii) $\dfrac{-2x^2 + 2}{(x^2 - x + 1)^2}$.

2 Let $y = x^{1/3}$, then $x = y^3$ so $\dfrac{dy}{dx} = \dfrac{1}{\dfrac{dx}{dy}} = \dfrac{1}{3y^2} = \dfrac{1}{3}x^{-2/3}$.

3 (i) $\dfrac{x}{\sqrt{x^2 + 1}}$, (ii) $\dfrac{1 - 3x^2}{2(1 + x^2)^2\sqrt{x}}$, (iii) $\dfrac{1}{\sqrt{(1 + x)(1 - x)^3}}$.

Exercise 2.3.1

1 (i) $-3 \sin 3x$, (ii) $2 \sin x \cos x = \sin 2x$, (iii) $2x \cos(x^2)$, (iv) $\sec^2 x$.

2 Let $y = \tan^{-1} x$, then $x = \tan y$, so $\dfrac{dy}{dx} = \dfrac{1}{\sec^2 y} = \dfrac{1}{1 + \tan^2 y} = \dfrac{1}{1 + x^2}$.

Exercises 2.4.1

1 $\dfrac{d}{dx} a^x = \dfrac{d}{dx} e^{x \log_e a} = e^{x \log_e a} \log_e a = a^x \log_e a$.

2 $\dfrac{d}{dx} \sinh x = \dfrac{d}{dx} \dfrac{1}{2}(e^x - e^{-x}) = \dfrac{1}{2}(e^x + e^{-x}) = \cosh x$. Similarly,

$\dfrac{d}{dx} \cosh x = \sinh x$. Write $\tanh x = \dfrac{\sinh x}{\cosh x}$ and use the quotient rule.

3 Let $y = \sinh^{-1} \dfrac{x}{a}$, then $x = a \sinh y$ and $\dfrac{dy}{dx} = \dfrac{1}{a \cosh y} = \dfrac{1}{a\sqrt{1 + \sinh^2 y}} =$

$\dfrac{1}{\sqrt{a^2 + x^2}}$.

4 Let $y = \log_a x$, then $x = a^y$, so $\dfrac{dy}{dx} = \dfrac{1}{a^y \log_e a} = \dfrac{1}{x \log_e a}$.

5 (i) $3^{\sin x} \ln 3 \cdot \cos x$, (ii) $-3^x \log_e 3 \cdot \sin(3^x)$.

Exercises 2.5.1

1 (i) $5x^4 - 3x^2$, $20x^3 - 6x$, $60x^2 - 6$; (ii) $-\dfrac{1}{x^2}$, $\dfrac{2}{x^3}$, $-\dfrac{6}{x^4}$;

(iii) $x \cos x + \sin x$, $2 \cos x - x \sin x$, $-3 \sin x - x \cos x$;
(iv) $\sinh x$, $\cosh x$, $\sinh x$.

2 (i) $e^x(316 \cos 3x + 12 \sin 3x)$, (ii) $e^x(x^3 + 30x^2 + 270x + 720)$.

Exercises 2.6.1

1 Let $f(x) = \cos x$, then $f'(x) = -\sin x$, $f''(x) = -\cos x$, $f^{(3)}(x) = \sin x$, $f^{(4)}(x) = \cos x$ and these repeat every four. Thus $f(0) = 1$, $f'(0) = 0$, $f''(0) = -1$, $f^{(3)}(0) = 0$ and these values repeat in fours. When these values are put into $a_n = (f^{(n)}(0))/n!$, and these in turn are substituted into $f(x) = a_0 + a_1 x + a_2 x^2 + \cdots$, we obtain the required series.

2 For e^x, $\dfrac{d}{dx} \dfrac{x^n}{n!} = \dfrac{x^{n-1}}{(n-1)!}$, so the derivative of each term gives the previous

term and $\dfrac{d}{dx} e^x = e^x$. For $\cos x$, $\dfrac{d}{dx} (-1)^n \dfrac{x^{2n}}{(2n)!} = (-1)^n \dfrac{x^{2n-1}}{(2n-1)!}$, which

is minus the $(n-1)$st term in the expansion of $\sin x$. Thus $\dfrac{d}{dx} \cos x = -\sin x$.

sin, cosh and sinh all follow the same pattern as cos.

Miscellaneous Exercises 2

1 $\dfrac{dl}{dT} = 0.000\,012 + 0.000\,000\,22T = 0.000\,034$ m/°C when $T = 100\,°C$.

2 Velocity $= \dfrac{ds}{dt} = 20 - 10t = -980$ ms^{-1} when $t = 100$ s.

3 Rate of change of period $= \dfrac{dT}{dl} = \dfrac{\pi}{\sqrt{gl}} \approx \dfrac{\pi}{\sqrt{9.81}} \approx 1.003$ s m^{-1} when $l = 1$ m.

4 $\dfrac{dV}{dt} = \dfrac{dx}{dt} \dfrac{dV}{dx} = \pi x^2 \dfrac{dx}{dt}$, so $\dfrac{dx}{dt} = \dfrac{1}{\pi x^2} \dfrac{dV}{dt} = \dfrac{10}{\pi 10^2} \approx 0.032$ cm s^{-1} when

$x = 10$ cm and $\dfrac{dV}{dt} = 10$ cm^3 s^{-1}.

5 At time $t > 1$ h, the distance between the ships is

$s = \sqrt{(12t)^2 + 16^2(t-1)^2} = 4\sqrt{9t^2 + 16(t-1)^2}$. Thus

$\dfrac{ds}{dt} = 4 \dfrac{9t + 16(t-1)}{\sqrt{9t^2 + 16(t-1)^2}} \approx 19.8$ knots at $t = 4$ h.

6 Differentiating the formula implicitly with respect to t, we find

$$-\frac{1}{u^2}\frac{du}{dt} - \frac{1}{v^2}\frac{dv}{dt} = 0$$

so

$$\frac{dv}{dt} = -\frac{v^2}{u^2}\frac{du}{dt}$$

When

$$u = 5 \text{ cm}, \frac{du}{dt} = -6 \text{ cm s}^{-1}, f = 4 \text{ cm}$$

so

$$\frac{1}{5} + \frac{1}{v} = \frac{1}{4} \Rightarrow v = 20 \text{ cm}$$

Then

$$\frac{dv}{dt} = -\frac{20^2}{5^2}(-6) = 96 \text{ cm s}^{-1}$$

7

$$H = \frac{M}{2l}\left[\frac{1}{(x-l)^2} - \frac{1}{(x+l)^2}\right]$$

$$= \frac{M}{2lx^2}\left[\left(1-\frac{l}{x}\right)^{-2} - \left(1+\frac{l}{x}\right)^{-2}\right]$$

$$= \frac{M}{2lx^2}\left[1 + 2\frac{l}{x} + 3\frac{l^2}{x^2} + \cdots - \left(1 - 2\frac{l}{x} + 3\frac{l^2}{x^2} + \cdots\right)\right]$$

$$= \frac{2M}{x^3} + \text{smaller terms}$$

3 INTEGRATION

3.1 AREA AND DEFINITE INTEGRALS

Suppose that we wish to compute the area of the region bounded by the lines $y = 0$, $y = k$, $x = a$ and $x = b$, where k, a and b are constants. This region is shown shaded in Figure 3.1, and it is clear that we can evaluate its area as $k(b - a)$ simply by using the rule for a rectangle.

Now consider the problem of evaluating the area of the region, shown in Figure 3.2, bounded by $y = 0$, $y = f(x)$, $x = a$ and $x = b$. Here there is no simple way of finding the area. One possibility of finding an approximation to the area is to divide it up into a large number, n, of rectangles $R_1, R_2, \ldots, R_n$ having intervals $I_1, I_2, \ldots, I_n$ of small length $\delta x = (b - a)/n$ as bases (see Figures 3.3 and 3.4). For the height of the rectangle R_i we take $f(x_i)$, the value of f at some point x_i in the interval I_i. The total area of the rectangles is $\sum_{i=1}^{n} f(x_i)\delta x$*. The smaller δx (and, hence, the larger n) the closer we

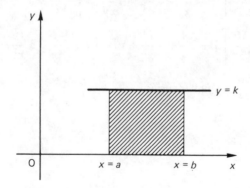

Fig. 3.1

---

*$\sum_{i=m}^{n} f(x_i)$ (read as the sum from m to n of $f(x_i)$) is a shorthand way of writing the sum

$$f(x_m) + f(x_{m+1}) + f(x_{m+2}) + \cdots + f(x_n).$$

Here i is the summation variable, which ranges consecutively through the integers, starting at m and finishing at n.

48

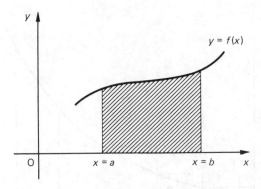

Fig. 3.2

should expect this sum to approximate the required area. If this sum has a limit as $\delta x \to 0$, we write it as $\int_a^b f(x)\mathrm{d}x$. This is called the *definite integral* of f from a to b, the numbers a and b are called the *limits of integration*, $[a, b]$ is called the *interval of integration* and $f(x)$ is called the *integrand*. We shall see that definite integrals are not normally calculated by this limiting process, but we illustrate the idea with an example.

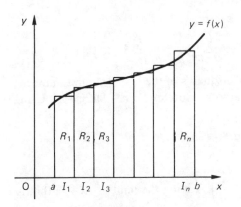

Fig. 3.3

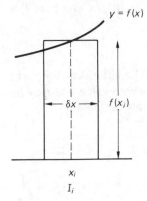

Fig. 3.4

Example 3.1.1

Calculate $\int_0^1 x^2 \, \mathrm{d}x$.

We divide the interval of integration into n subintervals $I_i = \left[\dfrac{i-1}{n}, \dfrac{i}{n}\right]$,

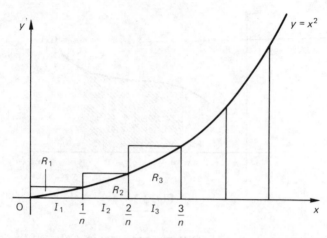

Fig. 3.5

each of length $\delta x = \dfrac{1}{n}$, as shown in Figure 3.5, and set $x_i = \dfrac{i}{n}$ (corresponding to the end of I_i), so that $f(x_i) = \dfrac{i^2}{n^2}$. The total area of the n rectangles R_i of height $f(x_i)$ on base I_i is

$$A_n = \sum_{i=1}^{n} f(x_i)\delta x = \sum_{i=1}^{n} \frac{i^2}{n^2}\frac{1}{n} = \frac{1}{n^3} \sum_{i=1}^{n} i^2$$

The latter sum is just the sum of the squares of the first n natural numbers, whose value you may have come across before as $\frac{1}{6}n(n+1)(2n+1)$. Using this, we obtain

$$A_n = \frac{(n+1)(2n+1)}{6n^2}$$

which we rewrite as $\dfrac{1}{6}\left(1 + \dfrac{1}{n}\right)\left(2 + \dfrac{1}{n}\right)$. Taking the limit as $\delta x \to 0$, that is, as $n \to \infty$, we obtain the value $\frac{1}{3}$. Thus, $\int_a^b x^2 \, dx = \frac{1}{3}$. This method of calculating areas was first used by Archimedes and so predates the development of calculus by many centuries.

Notice that if, in the above, $f(x_i)$ is negative, then the rectangle R_i will be below the x axis. Its area is $-f(x_i)\delta x$, since areas are normally taken as positive. For integrals, however, we adopt the convention that any contribution from below the axis will be negative; this is the case if we take $f(x_i)\delta x$ as the contribution of the rectangle R_i to the area A_n. This means that $A_n = \sum_{i=1}^{n} f(x_i)\delta x$ however the sign of f varies in the interval of integration.

50

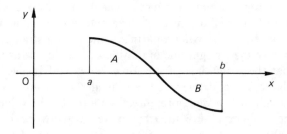

Fig. 3.6

For a function which crosses the x axis between a and b, such as that sketched in Figure 3.6, $\int_a^b f(x)\mathrm{d}x = A - B$, where A and B are the areas indicated.

We shall postpone further development for the moment, while we consider how integrals might arise in practice, other than as a means of calculating area.

EXERCISE 3.1.1

Use the above method to compute $\int_0^1 x\,\mathrm{d}x$. (Note that $\sum_{i=1}^n i = \frac{1}{2}n(n+1)$.)

3.2 SPEED AND DISTANCE, FORCE AND WORK

Consider a particle moving in a straight line whose speed at time t is given by $v(t)$, and suppose that we require to know the distance the particle travels between the times $t = a$ and $t = b$. As with area, the easiest case is that of constant speed, $v(t) = k$, say, where the required distance is $k(b-a)$ in the appropriate units (metres if t is in seconds and v in metres per second). For more general functions v we can calculate the distance, using essentially the same method as in Section 3.1. We divide the time interval $[a, b]$ into n subintervals, each of length $\delta t = (b-a)/n$. In the ith such subinterval we assume that the speed is constant, with value $v(t_i)$, where t_i is a value of t in this subinterval. The distance travelled by the particle in this subinterval of time is approximately $v(t_i)\delta t$, so that the total distance is approximately $\sum_{i=1}^n v(t_i)\delta t$. Letting the subintervals shrink to zero, that is, letting $\delta t \to 0$, the limit of the sum is $\int_a^b v(t)\mathrm{d}t$, just as in the earlier case of calculating the area. So any method we can devise for the evaluation of areas will also apply to the calculation of distance from a knowledge of speed as a function of time.

Now we consider the problem of calculating the work done by a variable force in moving a particle along a straight line from one point to another. (The more general problem where the path taken by the particle is not a straight line will be considered in a later chapter.) The simplest case is where the force is constant. In this case the work done is given by ks, where k is the magnitude of the force and s is the distance between the two points.

Again the appropriate units for measuring the work done will depend on those used for force and distance. For example, a joule is the work done by a force of 1 newton in moving a particle 1 metre in a straight line.

Now think of the straight line of motion of the particle as the x axis and suppose that the magnitude of the force varies and is given by $f(x)$. To calculate the total work done, we again divide $[a, b]$ into n subintervals of length δx within which we approximate the force by the value of f at some point in the subinterval. For the ith subinterval the work done is approximately $f(x_i)\delta x$, and for the whole interval $\sum_{i=1}^{n} f(x_i)\delta x$. Letting $\delta x \to 0$, we obtain the limit $\int_a^b f(x)\mathrm{d}x$. Once again any method of calculating areas will serve to calculate the answer.

We shall consider in the remainder of this chapter methods of evaluating integrals; they depend essentially on showing that integration is the inverse process to differentiation.

3.3 THE FUNDAMENTAL THEOREM OF CALCULUS

The speed–distance problem considered in Section 3.2 suggests the connection between integration and differentiation. There we were given the speed $v(t)$ at time t and calculated the distance moved in a time interval $[a, b]$ as $\int_a^b v(t)\mathrm{d}t$. Let us replace the limit b by x and write the distance at time x as $s(x)$, since it clearly depends on x. Thus, we have $s(x) = \int_a^x v(t)\mathrm{d}t$. However, if we had been given the problem the other way round, that is, given $s(t)$, we were required to find $v(t)$, we should have differentiated with respect to t to obtain $v(t) = (\mathrm{d}/\mathrm{d}t)s(t)$. This suggests that, to integrate a function f, all we must do is find a function, F, say, whose derivative is f. We now make this more explicit.

Consider a member f of our family of standard functions defined on $[a, b]$ and let $F(x) = \int_a^x f(t)\mathrm{d}t$, so that, for the function sketched in Figure 3.7, $F(x)$ is the area of the shaded region. We might describe $F(x)$ as the area up to x. Let us now calculate the derivative of this function F; since it is not a standard function, we must use first principles.

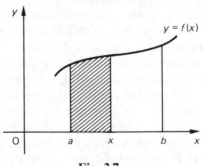

Fig. 3.7

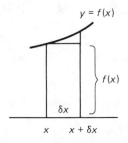

Fig. 3.8

The expression $F(x + \delta x) - F(x)$ represents the difference between the area up to $x + \delta x$ and the area up to x. It is thus the area between x and $x + \delta x$, which is illustrated in Figure 3.8. We can approximate it by the area $f(x)\delta x$ of the rectangle of height $f(x)$ based on $[x, x + \delta x]$. We now have

$$\frac{F(x + \delta x) - F(x)}{\delta x} \approx \frac{f(x)\delta x}{\delta x} = f(x)$$

As we reduce the size of δx, we expect the approximation to improve. In the limit as $\delta x \to 0$, the left-hand member becomes $F'(x)$, so that we have $F'(x) = f(x)$. This result, namely that $F(x) = \int_a^x f(x)\mathrm{d}x$ implies $F'(x) = f(x)$, is called the *Fundamental Theorem of Calculus*.

Now suppose that we can find *any* function p such that $p'(x) = f(x)$. Then $(F - p)'(x) = F'(x) - p'(x) = f(x) - f(x) = 0$. Thus, $F - p$ is a function whose derivative is zero and is therefore a constant, so that we can write $F(x) = p(x) + c$. Now $F(a)$ is the area up to a, and is clearly zero. Hence, $p(a) + c = 0$, giving $c = -p(a)$. Then

$$\int_a^b f(t)\mathrm{d}t = F(b) = p(b) + c = p(b) - p(a) \tag{1}$$

where p is any function satisfying $p'(x) = f(x)$. The function p is called a *primitive* or *indefinite integral* of f. We shall write $p(x) = \int f(x)\mathrm{d}x$; this differs from the notation in our derivation, where we used the variable t to the right of the integral sign. We need to use x in this context, however, since otherwise the variable x would not appear at all on the right-hand side. For example, x^2 is a primitive of $2x$, since $x^2 = \int 2x\,\mathrm{d}x$.

The problem of evaluating a definite integral is now reduced to finding a primitive p, substituting in the limits and subtracting. We use the notation $[p(x)]_a^b$ to denote $p(b) - p(a)$.

As we saw above, if p is a primitive of f, then so is $p + c$ for any constant c. We shall therefore include an arbitrary constant whenever an indefinite integral is evaluated. This constant will play a very important role later on, when we consider the solution of differential equations.

3.4 STANDARD INTEGRALS AND PROPERTIES OF INTEGRALS

The easiest functions to integrate are those which are derivatives of familiar functions. Examples are

$$\int x^n \, dx = \frac{x^{n+1}}{n+1} + c \qquad (n \neq -1)$$

$(n = 0$ gives $\int 1 \, dx = \int dx = x + c.)$

$$\int \sin x \, dx = -\cos x + c$$

$$\int e^x \, dx = e^x + c$$

$$\int \sec^2 x \, dx = \tan x + c$$

These appear together with many others in the list of standard integrals at the end of the chapter.

Another example is $\int dx/x = \ln x + c$. Note that the right-hand side is only defined when $x > 0$. If $x < 0$, then $\ln(-x)$ is defined and

$$\frac{d}{dx} \ln(-x) = \frac{1}{(-x)}(-1) = \frac{1}{x}$$

Hence

$$\int \frac{dx}{x} = \begin{cases} \ln x + c & \text{if } x > 0 \\ \ln(-x) + c & \text{if } x < 0 \end{cases}$$

or, more concisely,

$$\int \frac{dx}{x} = \ln |x| + c$$

Note that $\dfrac{1}{x}$ is undefined at $x = 0$, so that $\displaystyle\int_a^b \frac{dx}{x}$ is only defined if a and b have the same sign.

Properties of Integrals

In Chapter 2 we saw how to differentiate functions in the standard family by using rules for differentiating functions formed from simpler ones by addition, multiplication, composition and inversion. Unfortunately, there are few such rules for integration, but we shall 'reverse' the rules for differentiating products and compositions of functions to derive some techniques of integration in the next section.

54

The rules which we do have are

(1) $\int (f(x)+g(x))dx = \int f(x)dx + \int g(x)dx$
(2) $\int kf(x)dx = k\int f(x)dx$, where k is any constant
(3) $\int_a^b f(x)dx = -\int_b^a f(x)dx$
(4) $\int_a^c f(x)dx = \int_a^b f(x)dx + \int_b^c f(x)dx$, where $a<b<c$

Rules 1 and 2 can easily be checked by differentiation; the other two follow from Equation (1) of Section 3.3.

EXERCISES 3.4.1

1 Evaluate the following integrals:

(i) $\displaystyle\int_1^e \frac{dx}{x}$; (ii) $\displaystyle\int_{-e^2}^{-e} \frac{dx}{x}$; (iii) $\displaystyle\int_1^2 x\,dx$; (iv) $\displaystyle\int_0^\pi \cos x\,dx$; (v) $\displaystyle\int_0^\pi \sin x\,dx$.

2 Confirm the result of Exercise 3.1.1.
3 Integrate term by term the first five power series given before Exercises 2.6.1 and choose the constants of integration to make the series correct for $x = 0$.

3.5 INTEGRATION BY SUBSTITUTION

This method of integration is based on the chain rule for the differentiation of composite functions (see Section 2.2). Before looking at the method in general, we work an example in reverse to show how the method works.

Example 3.5.1

Find dy/dx when $y = \cos^5 x$.
Let $u = \cos x$, so that $y = u^5$. The chain rule gives

$$\frac{dy}{dx} = \frac{dy}{du}\frac{du}{dx} = 5u^4 \cdot (-\sin x) = -5\cos^4 x \sin x$$

The Fundamental Theorem now tells us that

$$\int -5\cos^4 x \sin x\,dx = \cos^5 x + c$$

Had we been given the integration problem in the first place, we should have to have spotted that, apart from the sign, sin is the derivative of cos, so that we could write the integral as

$$\int 5 \cos^4 x \, d(\cos x)$$

which makes it clear that it has come from differentiating a function (the fifth power) of a function (cos).

Now we look at the general form of the method. Let f be a function with primitive F, so that $F(u) = \int f(u)du$ and $F' = f$, and suppose that u is a function g of x. Then

$$\frac{d}{dx} F(g(x)) = \frac{d}{du} F(u) \frac{du}{dx}$$

$$= f(u) \cdot g'(x)$$

$$= f(g(x)) \cdot g'(x)$$

Thus,

$$\int f(g(x)) \cdot g'(x) dx = F(g(x)) = F(u) = \int f(u) du$$

which can be written as

$$\int f(g(x)) \frac{du}{dx} dx = \int f(u) du \qquad (1)$$

We have changed the integration with respect to x into one with respect to u, effectively by replacing $(du/dx)dx$ by du. We should remember, however, that these quantities should only be used in conjunction with an integral sign.

In our example above, with $u = \cos x$, so that $du/dx = -\sin x$, and $f(u) = u^4$, Equation (1) gives

$$\int \cos^4 x \cdot (-\sin x) dx = \int u^4 \, du = \frac{1}{5} u^5 + c = \frac{1}{5} \cos^5 x + c$$

Changing the sign throughout gives the answer.

We now look at some examples, starting with straightforward ones which we can immediately recognise as being in the form of Equation (1), and moving to harder ones where the substitution is not so obvious.

Examples 3.5.2

1. $\int x e^{x^2} dx$.

   With $u = x^2$ and $f(u) = e^u$, the integrand $x e^{x^2}$ becomes $\frac{1}{2} f(x^2)(du/dx)$.

Using Equation (1),

$$\int xe^{x^2}\,dx = \frac{1}{2}\int e^u\,\frac{du}{dx}\,dx = \frac{1}{2}\int e^u\,du = \frac{1}{2}e^u + c = \frac{1}{2}e^{x^2} + c$$

Now that we know how the method works, we make use of the remark following Equation (1) to simplify the solution. Thus, for the given integral we make the substitution $u = x^2$, $du = (du/dx)dx = 2x\,dx$, giving $x\,dx = \frac{1}{2}du$ and

$$\int xe^{x^2}\,dx = \int e^{x^2}x\,dx = \frac{1}{2}\int e^u\,du = \frac{1}{2}e^u + c = \frac{1}{2}e^{x^2} + c$$

2. $\int x^2\sin(x^3)dx$.

$\sin(x^3)$ is the function sin of the function x^3. (We should not call x^3 a function, since it is really the value of the cube function at x; however, we do not have a convenient and brief way of writing power functions, so, in common with most mathematicians, we use this abuse of notation. Thus, x^n can mean the nth power function or its value at x; the correct interpretation is usually clear from the context.) We thus use the substitution $u = x^3$, giving $du = (du/dx)dx = 3x^2\,dx$. Hence, $x^2\,dx = \frac{1}{3}du$ and

$$\int x^2\sin(x^3)dx = \int(\sin x^3)x^2\,dx = \frac{1}{3}\int\sin u\,du$$

$$= -\frac{1}{3}\cos u + c$$

$$= -\frac{1}{3}\cos(x^3) + c$$

3. $\int\tan x\,dx$.

This can be evaluated, using the method of substitution, as follows. First we replace $\tan x$ by $\sin x/\cos x$, then put $u = \cos x$, giving $du = -\sin x\,dx$. Thus,

$$\int\tan x\,dx = \int\frac{1}{\cos x}\sin x\,dx$$

$$= -\int\frac{1}{u}\,du$$

$$= -\ln|u| + c$$

$$= -\ln|\cos x| + c$$

$$= \ln|\sec x| + c$$

This is an example of a frequently occurring type of integral, namely $\int \frac{g'(x)}{g(x)} dx$. The required substitution here is $u = g(x)$, giving $du = g'(x)\, dx$, which transforms the integral into

$$\int \frac{1}{u}\, du = \ln|u| + c$$

Thus,

$$\int \frac{g'(x)dx}{g(x)} = \ln|g(x)| + c$$

4. $\int \dfrac{(2x+b)dx}{x^2+bx+c}$.

This is of the above form, with $g(x) = x^2 + bx + c$, so its value is $\ln|x^2 + bx + c| + k$.

5. $\int \cos(2x-1)dx$.

The substitution $u = 2x - 1$ gives $du = (du/dx)dx = 2\, dx$, so that $dx = \frac{1}{2}du$ and

$$\int \cos(2x-1)dx = \frac{1}{2}\int \cos u\, du$$

$$= \frac{1}{2}\sin u + c$$

$$= \frac{1}{2}\sin(2x-1) + c$$

This is a particular case of a function (cos) of a *linear* function $(2x-1)$. The general form of this is $\int f(ax+b)dx$, which the substitution $u = ax + b$, $du = a\, dx$, transforms into $\dfrac{1}{a}\int f(u)du$.

6. $\int \cos^2 x\, dx$.

Substituting for $\cos^2 x$ with the help of the trigonometric identity $\cos 2x = 2\cos^2 x - 1$, the integral becomes

$$\frac{1}{2}\int (1 + \cos 2x)dx = \frac{1}{2}x + \frac{1}{4}\sin 2x + c$$

EXERCISES 3.5.1

1 Use a suitable substitution to evaluate the following integrals:

(i) $\displaystyle\int 2x \cos(x^2)dx$; (ii) $\displaystyle\int x^3 e^{x^4}\, dx$; (iii) $\displaystyle\int \frac{\cos x}{\sin x} dx$;

(iv) $\displaystyle\int \frac{2x+3}{x^2+3x+4}\, dx$; (v) $\displaystyle\int \frac{x}{\sqrt{1-x^2}}\, dx$;

(vi) $\displaystyle\int \frac{x+1}{(x^2+2x)^3}\, dx$; (vii) $\displaystyle\int (3x+2)^4\, dx$; (viii) $\displaystyle\int e^{-5x}\, dx$;

(ix) $\displaystyle\int \cos 7x\, dx$.

2 Evaluate the following integrals:

  (i) $\int \sin 2x \sin 3x\, dx$; (ii) $\int \sin 2x \cos 3x\, dx$. (*Hint*: Use identities of the form

$$2 \sin mx \sin nx = \cos(m-n)x - \cos(m+n)x$$

$$\cos mx \cos nx = \cos(m+n)x + \cos(m-n)x$$

$$2 \sin mx \cos nx = \sin(m+n)x + \sin(m-n)x.)$$

3.6 INTEGRALS INVOLVING THE FACTOR $\sqrt{\pm a^2 \pm x^2}$

After evaluating a number of integrals, using the substitution techniques of the last section, the pattern should become quite familiar. We now turn to some less obvious candidates for the technique, using examples to illustrate how it can be applied.

Example 3.6.1

$\displaystyle\int \frac{dx}{a^2+x^2}$ (this is a standard integral).

Set $x = a \tan u$, giving $dx = \dfrac{dx}{du}\, du = a \sec^2 u\, du$ (notice that here the substitution is the opposite way round to those we have used before; we can think of this simply as an exchange of the variables x and u). Now $a^2 + x^2 = a^2(1 + \tan^2 u) = a^2 \sec^2 u$, so that

$$\int \frac{dx}{a^2+x^2} = \int \frac{a \sec^2 u}{a^2 \sec^2 u}\, du$$

$$= \frac{1}{a} \int du$$

$$= \frac{1}{a} u + c$$

$$= \frac{1}{a} \tan^{-1}\left(\frac{x}{a}\right) + c$$

Why did we pick this particular substitution? The reason is simply that it transformed the integral into one we could do. In evaluating integrals, we must try to imagine what substitution might help to simplify it. In this case the substitution $x = a \tan u$ enabled us to combine the two terms in the denominator to cancel, apart from the constant a, the numerator. Often we have to try a number of different substitutions before we find one that works.

Having identified a suitable substitution of a function of u for x in $\int f(x)dx$, the procedure to follow is:

(1) Work out dx/du in terms of u. Substitute for x in $f(x)$ and replace dx by $(dx/du)du$ to obtain the integral in the form $\int g(u)du$.
(2) Find an integral of g.
(3) Write the result in terms of x. This requires us to express u in terms of x and may well involve the use of inverse functions and trigonometric identities.

We now categorise the integrand according to the signs of a^2 and x^2:

(a) When the integrand contains $\sqrt{a^2 - x^2}$, we usually try the substitution $x = a \sin u$, since this expression becomes $\sqrt{a^2 - a^2 \sin^2 u} = a \cos u$ with the help of the identity $\cos^2 u + \sin^2 u = 1$. Rearranging the identity $\cosh^2 u - \sinh^2 u = 1$ into $\sinh^2 u = \cosh^2 u - 1$ and dividing through by $\cosh^2 u$ yields $\tanh^2 u = 1 - \text{sech}^2 u$; this suggests the alternative substitution $x = a \,\text{sech}\, u$, giving $\sqrt{a^2 - x^2} = a \tanh u$.
(b) When the integrand contains $\sqrt{a^2 + x^2}$, the first substitution to try is $x = a \tan u$, which gives $\sqrt{a^2 + x^2} = \sqrt{a^2 + a^2 \tan^2 u} = a \sec u$. Here we have made use of the identity $1 + \tan^2 u = \sec^2 u$, which we can obtain by dividing the identity $\cos^2 u + \sin^2 u = 1$ through by $\cos^2 u$. An alternative substitution in this case is $x = a \sinh u$, which reduces $\sqrt{a^2 + x^2}$ to $a \cosh u$.
(c) When the integrand contains $\sqrt{x^2 - a^2}$, the substitution $x = a \cosh u$ reduces the expression to $a \sinh u$, or the substitution $x = a \sec u$ reduces it to $a \tan u$.

For each form of integrand, we have suggested two possible substitutions. Hopefully, one of these will transform the integral into one which is reasonably easy to evaluate.

Examples 3.6.2

1. $\displaystyle \int \frac{dx}{x^2\sqrt{1 + x^2}}$.

Put $x = \tan u$, so that $dx = \sec^2 u \, du$ and $\sqrt{1 + x^2} = \sec u$. Then

$$\int \frac{dx}{x^2\sqrt{1 + x^2}} = \int \frac{\sec^2 u}{\tan^2 u \sec u}\, du = \int \frac{\cos u}{\sin^2 u}\, du = -\text{cosec}\, u + c$$

where the integration can be done using the further substitution $\sin u = t$. We complete the example by using the identity $\operatorname{cosec}^2 u = \cot^2 u + 1$ (obtained by dividing $1 = \cos^2 u + \sin^2 u$ by $\sin^2 u$) to obtain the value of the integral as $-\sqrt{1 + \dfrac{1}{x^2}} + c$.

2. $\int \sqrt{1 - x^2}\, dx$.

Put $x = \sin u$, $dx = \cos u\, du$, when the integral becomes $\int \cos^2 u\, du$. Here we use a suitable trigonometric identity to obtain the integral as

$$\int \frac{1}{2}(1 + \cos 2u)du = \frac{1}{2}u + \frac{1}{4}\sin 2u + c$$

$$= \frac{1}{2}(u + \sin u \cos u) + c$$

$$= \frac{1}{2}(\sin^{-1} x + x\sqrt{1 - x^2}) + c$$

where we have used the identity $\cos^2 u + \sin^2 u = 1$ to express $\cos u$ in terms of x.

3. $\int \dfrac{dx}{\sqrt{x^2 - 1}}$.

The substitution $x = \sec u$, $dx = \sec u \tan u$ gives

$$\int \frac{dx}{\sqrt{x^2 - 1}} = \int \sec u\, du$$

$$= 1n|\sec u + \tan u| + c$$

$$= 1n|x + \sqrt{x^2 - 1}| + c$$

using the identity $1 + \tan^2 u = \sec^2 u$.

Alternatively, putting $x = \cosh u$, $dx = \sinh u\, du$ transforms the integral into $\int du = u = \cosh^{-1} x + c$.

As we saw in Example 1.9.2, the two answers agree for $x > 1$; for $x < 1$, however, the integral does not exist.

4. $\int \sqrt{x^2 - 1}\, dx$, $x > 1$.

Put $x = \cosh u$, $dx = \sinh u\, du$, to obtain

$$\int \sqrt{x^2 - 1}\, dx = \int \sinh^2 u\, du$$

$$= \int \left(\frac{e^u - e^{-u}}{2}\right)^2 du$$

61

$$= \frac{1}{4} \int (e^{2u} - 2 + e^{-2u}) du$$

$$= \frac{1}{8} e^{2u} - \frac{1}{2} u - \frac{1}{8} e^{-2u} + c$$

$$= \frac{1}{2} \frac{1}{2} (e^u - e^{-u}) \frac{1}{2} (e^u + e^{-u}) - \frac{1}{2} u + c$$

$$= \frac{1}{2} \sinh u \cosh u - \frac{1}{2} u + c$$

$$= \frac{1}{2} (x\sqrt{x^2 - 1} - \cosh^{-1} x) + c$$

where we have used the identity $\cosh^2 u - \sinh^2 u = 1$ to express $\sinh u$ in terms of x. Here we chose to express $\sinh^2 u$ in terms of exponentials in order to integrate it. The substitution $\sinh^2 u = \frac{1}{2}(\cosh 2u - 1)$ would also have worked, but exponentials are particularly easy to integrate.

This integral can be used to obtain the result in Section 1.8 for the area of the shaded region in Figure 3.9.

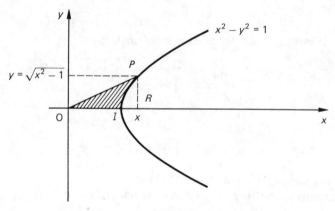

Fig. 3.9

The required area equals

$$\text{area triangle ORP} - \int_1^x \sqrt{t^2 - 1} \, dt$$

$$= \frac{1}{2} x\sqrt{x^2 - 1} - \frac{1}{2} \left[t\sqrt{t^2 - 1} - \cosh^{-1} t \right]_1^x$$

$$= \frac{1}{2} \cosh^{-1} x$$

Thus, if P has coordinates (cosh t, sinh t) the area of the 'sector' OIP of the hyperbola is $\frac{1}{2}t$.

We summarise suitable substitutions in the following table.

Integrand factor	Substitution
$\sqrt{a^2 - x^2}$	$x = a \sin u$ or $x = a \operatorname{sech} u$
$\sqrt{a^2 + x^2}$	$x = a \tan u$ or $x = a \sinh u$
$\sqrt{x^2 - a^2}$	$x = a \cosh u$ or $x = a \sec u$

Definite Integrals by Substitution

In the last example, although we used substitution to evaluate the integral, we expressed the final answer in terms of the original variable. It is usually more convenient instead to change the limits into values of the new variable. For the indefinite integral, the substitution $x = g(u)$, $dx = g'(u)du$ gives

$$\int f(x)dx = \int f(g(u))g'(u)du$$

Suppose now that we wish to evaluate the integral on the left with limits a and b. The corresponding values, c, d, say, of u are given as the solutions of $a = g(c)$ and $b = g(d)$. If g has an inverse, we have simply $c = g^{-1}(a)$ and $d = g^{-1}(b)$. In fact, since the integral in terms of u is only defined by the integral given in terms of x, u must be uniquely defined in terms of x, that is, g^{-1} must exist at all points of the interval of integration, so that $u = g^{-1}(x)$.

If f has primitive p, then

$$\int_a^b f(x)dx = p(b) - p(a)$$

$$= p(g(d)) - p(g(c))$$

$$= \int_c^d f(g(u))g'(u)du$$

where $u = d$ when $x = b$ and $u = c$ when $x = a$. In practice, definite integrals are often easier than indefinite integrals, because they do not require an explicit inverse function.

63

Example 3.6.3

Evaluate $\int_0^1 \sqrt{1-x^2}\,dx$.

As in Example 3.6.2.2, we make the substitution $x = \sin u$. When $x = 0$, $u = 0$ and when $x = 1$, $u = \pi/2$. We thus have

$$\int_0^1 \sqrt{1-x^2}\,dx = \int_0^{\pi/2} \cos^2 u\,du$$

$$= \frac{1}{2}\int_0^{\pi/2} (1 + \cos 2u)\,du$$

$$= [\tfrac{1}{2}u + \tfrac{1}{4}\sin 2u]_0^{\pi/2}$$

$$= \frac{\pi}{4}$$

This integral gives the area under the curve $y = \sqrt{1-x^2}$, that is, the area of the first quadrant of a circle of unit radius.

EXERCISES 3.6.1

1 Use a suitable substitution to evaluate the following integrals:

(i) $\int (1+x^2)^{-3/2}\,dx$; (ii) $\int x^2(x^2-4)^{-3/2}\,dx$; (iii) $\int_0^1 x^2\sqrt{1-x^2}\,dx$.

2 Show that if $g(x) = \sec x + \tan x$, then

$$\frac{g'(x)}{g(x)} = \sec x$$

Hence evaluate $\int \sec x\,dx$.

3.7 INTEGRATION BY PARTS

This method is based on the product rule for the differentiation of the product of two functions. Let u and v be functions of x. Integrating the rule

$$\frac{d}{dx}(uv) = u\frac{dv}{dx} + v\frac{du}{dx}$$

with respect to x gives

$$uv = \int u \frac{dv}{dx} \, dx + \int v \frac{du}{dx} \, dx$$

$$= \int u \, dv + \int v \, du$$

We rearrange this to obtain

$$\int u \, dv = uv - \int v \, du$$

This formula is useful if the right-hand integral can be evaluated directly or with further reduction. The trick lies in choosing the functions u and v to make the right-hand integral possible.

Examples 3.7.1

1. Evaluate $\int xe^x \, dx$.

   In this integral we choose $u = x$ because differentiation will make it into a constant. Thus, $du = dx$; we must now choose $dv = e^x \, dx$, giving $v = e^x$ (no constant of integration is required here). So

   $$\int xe^x \, dx = xe^x - \int e^x \, dx$$

   $$= xe^x - e^x + c$$

2. Evaluate $\int x^2 \sin x \, dx$.

   Put $u = x^2$ and $dv = \sin x \, dx$, giving $du = 2x \, dx$ and $v = -\cos x$, so that

   $$\int x^2 \sin x \, dx = -x^2 \cos x + 2 \int x \cos x \, dx \tag{1}$$

   This stage has reduced the power of x by one; another integration by parts will reduce it to a constant. Thus, putting $u = x$ and $dv = \cos x \, dx$ gives $du = dx$ and $v = \sin x$, so that

   $$\int x \cos x \, dx = x \sin x - \int \sin x \, dx$$

   $$= x \sin x + \cos x + c$$

   Substituting this into Equation (1), we obtain

   $$\int x^2 \sin x \, dx = -x^2 \cos x + 2x \sin x + 2 \cos x + c$$

The last two examples are particular cases with integrands of the form $x^n f(x)$, where $f(x) = \sin x$, $\cos x$ or e^x. Almost always, integrals of this kind are done by a succession of integrations by parts, in which each stage reduces the power of x by one. For powers of x greater than two, it is useful to construct a reduction formula, as in the following two examples.

3. Find a reduction formula for $I_n = \int x^n e^{2x} dx$ and use it to evaluate I_2.

For $n \geq 1$, put $u = x^n$ and $dv = e^{2x} dx$, so that $du = nx^{n-1}$ and $v = \frac{1}{2} e^{2x}$. We have, after integrating by parts,

$$I_n = \frac{1}{2} x^n e^{2x} - \frac{n}{2} \int x^{n-1} e^{2x} dx$$

$$= \frac{1}{2} x^n e^{2x} - \frac{n}{2} I_{n-1}$$

This is the required reduction formula; repeated applications enable us to reduce the integral to I_0, which we can evaluate directly. Thus,

$$I_2 = \frac{1}{2} x^2 e^{2x} - I_1$$

$$= \frac{1}{2} x^2 e^{2x} - \left(\frac{1}{2} x e^{2x} - \frac{1}{2} I_0 \right)$$

$$= \frac{1}{2} x^2 e^{2x} - \frac{1}{2} x e^{2x} + \frac{1}{2} \int e^{2x} dx$$

$$= \frac{1}{2} x^2 e^{2x} - \frac{1}{2} x e^{2x} + \frac{1}{4} e^{2x} + c$$

4. Find a reduction formula for $I_n = \int_0^{\pi/2} \cos^n x\, dx$ and use it to obtain the value of $\int_0^{\pi/2} \cos^8 x\, dx$.

We split $\cos^n x$ and integrate by parts:

$$I_n = \int_0^{\pi/2} \cos^{n-1} x \cdot \cos x\, dx$$

$$= \left[\cos^{n-1} x \cdot \sin x \right]_0^{\pi/2} + (n-1) \int_0^{\pi/2} \cos^{n-2} x \cdot \sin^2 x\, dx$$

and, replacing $\sin^2 x$ by $1 - \cos^2 x$, this reduces to

$$I_n = (n-1) I_{n-2} - (n-1) I_n$$

Solving for I_n, we obtain the reduction formula

$$I_n = \frac{n-1}{n} I_{n-2}$$

For the case when $n = 8$, repeated application of the reduction formula yields

$$I_8 = \frac{7\,5\,3\,1}{8\,6\,4\,2} \int_0^{\pi/2} dx$$

$$= \frac{35\pi}{256}$$

5. Find reduction formulae for $I_{m,n} = \int_0^{\pi/2} \sin^m x \cos^n x \, dx$.

   If $n = 1$, the substitution $u = \sin x$ transforms the integral to

$$I_{m,1} = \int_0^1 u^m \, du = \left[\frac{u^{m+1}}{m+1} \right]_0^1 = \frac{1}{m+1}$$

Similarly,

$$I_{1,n} = \frac{1}{n+1}$$

For $m > 1$, $n \geq 0$, write

$$I_{m,n} = \int_0^{\pi/2} \sin^{m-1} x \cos^n x \sin x \, dx$$

which, after integration by parts, yields

$$I_{m,n} = \left[-\frac{\sin^{m-1} x \cos^{n+1} x}{n+1} \right]_0^{\pi/2} + \frac{m-1}{n+1} \int_0^{\pi/2} \sin^{m-2} x \cos^{n+2} x \, dx$$

$$= \frac{m-1}{n+1} I_{m-2,n+2} = \frac{m-1}{n+1} I_{m-2,n} - \frac{m-1}{n+1} I_{m,n}$$

(putting $\cos^2 x = 1 - \sin^2 x$)

Solving for $I_{m,n}$, we find

$$I_{m,n} = \frac{m-1}{m+n} I_{m-2,n}$$

Similarly,

$$I_{m,n} = \frac{n-1}{m+1} I_{m+2,n-2} = \frac{n-1}{m+n} I_{m,n-2}$$

These reduction formulae can be used to find, for example,

$$I_{6,4} = \frac{5}{10} I_{4,4} = \frac{5}{10} \frac{3}{8} I_{2,4} = \frac{5}{10} \frac{3}{8} \frac{3}{6} I_{2,2} = \frac{5}{10} \frac{3}{8} \frac{3}{6} \frac{1}{4} I_{2,0}$$

$$= \frac{5}{10} \frac{3}{8} \frac{3}{6} \frac{1}{4} \frac{1}{2} I_{0,0} = \frac{5}{10} \frac{3}{8} \frac{3}{6} \frac{1}{4} \frac{1}{2} \frac{\pi}{2} = \frac{3\pi}{512}$$

Another type of integral which can be evaluated, using integration by parts, has the form $I = \int e^{ax} \sin bx \, dx$ or $\int e^{ax} \cos bx \, dx$.

6. Evaluate $I = \int e^x \sin x \, dx$.

Put $u = \sin x$ and $dv = e^x \, dx$, giving $du = \cos x \, dx$ and $v = e^x$. We now have

$$I = \int e^x \sin x \, dx = e^x \sin x - \int e^x \cos x \, dx \qquad (1)$$

To evaluate the right-hand integral, we put $u = \cos x$ and $dv = e^x \, dx$, giving $du = -\sin x \, dx$ and $v = e^x$, so

$$\int e^x \cos x \, dx = e^x \cos x + \int e^x \sin x \, dx$$

$$= e^x \cos x + I$$

Putting this into Equation (1), we obtain

$$I = e^x(\sin x - \cos x) - I$$

Solving for I and introducing an arbitrary constant gives

$$I = \tfrac{1}{2}(\sin x - \cos x)e^x + c$$

Similar techniques can be used for integrals involving sinh and cosh rather than sin and cos; alternatively, sinh and cosh can be expressed in terms of the exponential function.

The functions we have integrated by parts so far have all been products of functions. Unfortunately, not all products can be so integrated. However, the method can be used for a straightforward integral, $\int f(x)dx$ say, simply by taking $u = f(x)$ and $dv = dx$.

---

7. Evaluate $\int \ln x \, dx$.

Put $u = \ln x$ and $dv = dx$, so that $du = dx/x$ and $v = x$. We have

$$\int \ln x \, dx = x \ln x - \int x \, \frac{dx}{x}$$

$$= x \ln x - x + c$$

To evaluate a definite integral, we simply insert the limits in the usual way as each integral is completed.

8. Evaluate $\int_1^2 \ln x \, dx$.

$$\int_1^2 \ln x \, dx = [x \ln x]_1^2 - \int_1^2 x \, \frac{dx}{x}$$

$$= 2 \ln 2 - [x]_1^2$$

$$= 2 \ln 2 - 1$$

EXERCISES 3.7.1

1 Evaluate the following integrals:

(i) $\int x \sin x \, dx$; (ii) $\int x^2 e^{3x} \, dx$; (iii) $\int x^4 \ln x \, dx$;
(iv) $\int e^{2x} \cos 3x \, dx$; (v) $\int \sec^3 x \, dx$.

(*Hint*: For (v) take $u = \sec x$, $dv = \sec^2 x \, dx$, then use $\tan^2 x = \sec^2 x - 1$.)

2 For an integer $n > 0$, let $I_n = \int \sin^n x \, dx$. Integrate by parts, using $u = \sin^{n-1} x$ and $dv = \sin x \, dx$ to show that if $n > 2$, then $nI_n = (n-1)I_{n-2} - \sin^{n-1} x \cos x$. (*Hint*: Replace $\cos^2 x$ by $1 - \sin^2 x$.) Find I_3 and I_4.

3 Find a reduction formula for evaluating $\int_0^{\pi/2} \sin^n x \, dx$.

3.8 PARTIAL FRACTIONS

Recall that we can add fractions, as in the following example:

$$\frac{2}{x-1} + \frac{3}{x+3} = \frac{2(x+3)}{(x-1)(x+3)} + \frac{3(x-1)}{(x+3)(x-1)} = \frac{5x+3}{(x-1)(x+3)}$$

Suppose, however, that the problem is turned round, so that we are given the last expression and wish to express it as two separate fractions. Since we have just seen the addition, it is easy to write

$$\frac{5x+3}{(x-1)(x+3)} = \frac{2}{x-1} + \frac{3}{x+3}$$

Each term on the right-hand side is called a *partial fraction* and we have expressed the left-hand side in terms of its partial fractions. If this can be done for all rational functions, it will provide a way of integrating them, since the partial fractions are easy to integrate. We now investigate how we can split rational functions into partial fractions, starting with some easy examples before we look at the general case.

Example 3.8.1

Express $(3x + 3)/(x^2 + x - 2)$ in terms of partial fractions.

We first note that $x^2 + x - 2$ factorises into $(x - 1)(x + 2)$. Now assume that the splitting into partial fractions can be done; specifically, assume that there are constants A and B, so that

$$\frac{3x + 3}{x^2 + x - 2} = \frac{A}{x - 1} + \frac{B}{x + 2}$$

is an identity, that is, it is true for all values of x. Multiplying through by $(x - 1)(x + 2)$ gives us the identity

$$3x + 3 = A(x + 2) + B(x - 1) \tag{1}$$

$$= (A + B)x + (2A - B)$$

Since this is an identity, we must have

$$A + B = 3 \quad \text{(the coefficient of } x\text{)}$$

$$2A - B = 3 \quad \text{(the constant term)}$$

The solution of these simultaneous equations is $A = 2, B = 1$, and so we have

$$\frac{3x + 3}{x^2 + x - 2} = \frac{2}{x - 1} + \frac{1}{x + 2}$$

Values of A and B can always be found in this way, but there is an easier way which avoids having to solve simultaneous equations. After obtaining Equation (1) as before, we can choose any value we like for x, since the identity must be true for all values of x. We choose values which make our calculations as easy as possible. Thus, taking $x = 1$ makes the coefficient of B in Equation (1) vanish, giving us directly $A = 2$. Similarly, putting $x = -2$ gives $B = 1$. We now look at a more difficult example.

Example 3.8.2

Evaluate $\displaystyle\int \frac{3x^3 + x^2 - 2x + 3}{x^4 - x^2 - 2x + 2} \, dx.$

Our first task is to factorise the denominator. The most straightforward way to start this is to try to find values of x which make the denominator vanish; if a is such a value of x, then $x - a$ is a factor of the denominator. By experimenting with a few integer values of x, we soon see that $x - 1$ is a factor in our present example. Divide this out, using long division as in the division of numbers:

$$
\begin{array}{r}
x^3 + x^2 + 0x - 2 \\
x - 1 \overline{\smash{\big)}\ x^4 - 0x^3 - x^2 - 2x + 2} \\
\underline{x^4 -\ x^3} \\
x^3 - x^2 \\
\underline{x^3 - x^2} \\
0 - 2x + 2 \\
\underline{-2x + 2} \\
0
\end{array}
$$

Putting $x = 1$ makes $x^3 + x^2 - 2$ vanish, so that $x - 1$ is a factor of this also. Dividing this factor out leaves the remaining factor $x^2 + 2x + 2$, which has no real factors. Altogether, then, we have

$$x^4 - x^2 - 2x + 2 = (x - 1)(x - 1)(x^2 + 2x + 2)$$

Compared with our simple example above, there are two new features: first, we have a *repeated* factor of $x - 1$, and second, we have a quadratic factor $x^2 + 2x + 2$. We accommodate these features by splitting into partial fractions in the following way:

$$\frac{3x^3 + x^2 - 2x + 3}{x^4 - x^2 - 2x + 2} = \frac{A}{x - 1} + \frac{B}{(x - 1)^2} + \frac{Cx + D}{x^2 + 2x + 2}$$

Wherever a quadratic denominator occurs, it must have a linear numerator. Although this does not seem to occur here, we can see that it in fact does, by combining the first two terms on the right to obtain

$$\frac{A(x - 1) + B}{(x - 1)^2}$$

We now follow the same procedure as before, starting by multiplying through by $(x - 1)^2(x^2 + 2x + 2)$, to obtain

$$
\begin{aligned}
3x^3 + x^2 - 2x + 3 = {} & A(x - 1)(x^2 + 2x + 2) + B(x^2 + 2x + 2) \\
& + (Cx + D)(x - 1)^2 \hspace{3em} (1) \\
= {} & (A + C)x^3 + (A + B - 2C + D)x^2 + (2B + C - 2D)x \\
& + (-2A + 2B + D)
\end{aligned}
$$

We can make this an identity by choosing the parameters so that the coefficients of powers of x are the same on each side. This yields

$$
\begin{array}{llll}
A & +\ C & =\ 3 & \text{(coefficient of } x^3) \\
A + & B - 2C + & D = \ 1 & \text{(coefficient of } x^2) \\
& 2B + \ C - 2D = -2 & & \text{(coefficient of } x) \\
-2A + 2B & +\ D = \ 3 & & \text{(constant)}
\end{array}
$$

71

This set of equations can be solved in the usual way, to obtain

$$A = 1, \; B = 1, \; C = 2, \; D = 3$$

In our simple example above, we saw that the solution of two simultaneous equations could be avoided by using judiciously chosen values of x. This would be even more beneficial here, where we ended up having to solve four simultaneous equations. Putting $x = 1$ in Equation (1) enables us to find $B = 1$ directly, but there is no other useful value of x to choose. Since Equation (1) is an identity, differentiating it will produce another identity, which, after rearranging, becomes

$$9x^2 + 2x - 2 = Ax(3x + 2) + 2B(x + 1) + C(x - 1)(3x - 1) + 2D(x - 1)$$

Putting $x = 1$ in this gives the equation

$$9 = 5A + 4B$$

from which we find, by putting in the value $B = 1$ already found, that $A = 1$. Differentiating again, we obtain

$$18x + 2 = 2A(3x + 1) + 2B + 2C(3x - 2) + 2D$$

Putting $x = 2/3$ yields the equation

$$14 = 6A + 2B + 2D$$

which gives $D = 3$. A final differentiation gives

$$18 = 6A + 6C$$

from which we obtain $C = 2$.

You may find this just as difficult as solving simultaneous equations, so you should pick whichever method you find easier.

Our integral can now be written as

$$\int \frac{dx}{x - 1} + \int \frac{dx}{(x - 1)^2} + \int \frac{(2x + 3)dx}{x^2 + 2x + 2}$$

The first two integrals can be evaluated by the method of substitution and we shall shortly be looking at how to evaluate the third.

We have looked at some examples which indicate how we might integrate some rational functions; in the next section, we look at the problem of integrating a general rational function.

EXERCISES 3.8.1

1 Put into partial fractions and integrate:

(i) $\dfrac{1}{(x+1)(x+2)}$; (ii) $\dfrac{x+1}{x(x+2)}$; (iii) $\dfrac{1}{x^2-a^2}$; (iv) $\dfrac{x-2}{x^2-4x+3}$;

(v) $\dfrac{x}{(x-2)^2}$.

2 Evaluate the following integrals:

(i) $\displaystyle\int \frac{dx}{(x+1)(x-2)(x-3)}$;

(ii) $\displaystyle\int \frac{(2x^2+5x-4)}{(x-1)^3}\,dx.$

3.9 SYSTEMATIC INTEGRATION OF RATIONAL FUNCTIONS

We consider the function f/g, where

$$f(x) = a_m x^m + a_{m-1} x^{m-1} + \cdots + a_0$$

and

$$g(x) = b_n x^n + b_{n-1} x^{n-1} + \cdots + b_0$$

We can always make $b_n = 1$ by dividing top and bottom by b_n. If $m \geq n$, we divide $f(x)$ by $g(x)$, to obtain

$$\frac{f(x)}{g(x)} = q(x) + \frac{r(x)}{g(x)}$$

where $q(x)$ is the quotient polynomial and $r(x)$ is the remainder polynomial, having degree less than n. We can easily integrate $q(x)$, so we have reduced the problem to that of integrating a rational function in which the degree of the numerator is less than that of the denominator. We illustrate the method of division by an example.

Example 3.9.1

Express $\dfrac{x^3 + 3x^2 + 6x + 5}{x^2 + 2x + 2}$ as the sum of a polynomial and a rational function with smaller degree in its numerator than its denominator.

As in Example 3.8.2, we use long division:

$$
\begin{array}{r}
x+1 \\
x^2+2x+2\,\overline{\smash{\big)}\,x^3+3x^2+6x+5} \\
\underline{x^3+2x^2+2x} \\
x^2+4x+5 \\
\underline{x^2+2x+2} \\
2x+3
\end{array}
$$

which shows that

$$
\frac{x^3+3x^2+6x+5}{x^2+2x+2}=x+1+\frac{2x+3}{x^2+2x+2}
$$

We now consider the splitting into partial fractions of $\dfrac{f(x)}{g(x)}$, where f has lower degree than g. The method depends on the fact that $g(x)$ can be factorised into factors which are either linear (of the form $x-a$) or irreducible quadratic (of the form x^2+bx+c with $b^2<4c$), and goes as follows:

(a) If $g(x)$ has a factor $(x-a)^r$, $r\geq 1$, then the partial fraction splitting should contain the terms

$$
\frac{A_1}{(x-a)}+\frac{A_2}{(x-a)^2}+\cdots+\frac{A_r}{(x-a)^r}
$$

After multiplication through by $g(x)$, the constants $A_1, A_2, \ldots, A_r$ can be found by putting $x=a$, differentiating, putting $x=a$ again, and so on.

(b) If $g(x)$ has a factor $(x^2+bx+c)^s$, where x^2+bx+c is irreducible and $s\geq 1$, then the partial fraction splitting must contain the terms

$$
\frac{C_1x+D_1}{x^2+bx+c}+\frac{C_2x+D_2}{(x^2+bx+c)^2}+\cdots+\frac{C_sx+D_s}{(x^2+bx+c)^s}
$$

After finding the constants appearing in (a), the constants $C_1, D_1, \ldots, C_s$, D_s can be found by comparing coefficients of x.

It turns out in all cases that the total number of constants to be found is equal to the degree of $g(x)$.

The problem of integrating general rational functions is thus reduced to that of integrating rational functions of the form

type (i): $(x-a)^{-n}$

type (ii) $\dfrac{Cx+D}{(x^2+bx+c)^n}$, where x^2+bx+c is irreducible

The substitution $u=x-a$ reduces type (i) to a standard integral. Before investigating the general case (ii), we look at two special cases, starting with

type (iii): $(x^2+a^2)^{-n}$

74

Here the substitution $x = a \tan u$ transforms the integrand to a power of $\cos u$, which we can integrate with the help of a reduction formula.

We can use the technique of completing the square to extend type (iii) to include integrals of the type

$$\int (x^2 + bx + c)^{-n}\, dx$$

where $x^2 + bx + c$ is irreducible. (If the zeros are real, we can use linear factors.) To do this we simply write

$$x^2 + bx + c = (x + \tfrac{1}{2}b)^2 + a^2$$

where $a = \sqrt{c - \tfrac{1}{4}b^2}$ is real, since $b^2 < 4c$ because $x^2 + bx + c$ is irreducible. Making the substitution $u = x + b/2$ now gives

$$\int (x^2 + bx + c)^{-n}\, dx = \int (u^2 + a^2)^{-n}\, du$$

which is of type (iii).

The second special case of type (ii) is

type (iv): $\dfrac{2x + b}{(x^2 + bx + c)^n}$

in which the numerator is the derivative of the factor in the denominator. The substitution $u = x^2 + bx + c$ reduces this to a standard integral.

We now show how to express type (ii) integrals in terms of type (iii) and type (iv) integrals by means of an example.

Example 3.9.2

Evaluate

$$I = \int \frac{(4x - 1)dx}{(x^2 + 2x + 2)^5}$$

Since $2^2 < 4 \times 2$, the denominator is irreducible, so this is a type (ii) integral. The derivative of $x^2 + 2x + 2$ is $2x + 2$ and since the numerator contains $4x$, we take 2 out before splitting the integral into the two desired parts:

$$I = 2 \int \frac{(2x + 2 - 5/2)}{(x^2 + 2x + 2)^5}\, dx$$

$$= 2 \int \frac{(2x + 2)dx}{(x^2 + 2x + 2)^5} - 5 \int \frac{dx}{((x + 1)^2 + 1)^5}$$

Here the first member is of type (iv), while the second is of type (iii), and

75

so we have written its denominator in the form of a completed square. The substitution $u = x^2 + 2x + 2$ reduces the first integral to

$$2 \int u^{-5} \, du = -\frac{1}{2} u^{-4} + C$$

The substitution $t = x + 1$ reduces the second integral to

$$5 \int (t^2 + 1)^{-5} \, dt$$

and the substitution $t = \tan u$ reduces this in turn to

$$5 \int \cos^8 u \, du$$

whose evaluation we leave as an exercise.

EXERCISES 3.9.1

1 Evaluate the integrals:

(i) $\displaystyle\int \frac{dx}{x^2 - 2x - 1}$; (ii) $\displaystyle\int \frac{dx}{(x^2 + 2x + 5)^2}$; (iii) $\displaystyle\int \frac{4x + 7}{x^2 + 10x + 29} \, dx.$

2 Evaluate the integral:

$$\int \frac{x^4 + 3x^3 + 10x^2 - 6x + 30}{x^2 + 4x + 13} \, dx.$$

3 Evaluate the integral

$$\int \frac{2 - 6x}{x^4 - 1} \, dx.$$

4 Evaluate

$$\int \frac{e^{2x} - 1}{e^x + e^{3x}} \, dx.$$

(*Hint*: Put $u = e^x$.)

5 Use the substitution $u = \sqrt{x - 1}$ to find

$$\int \frac{\sqrt{x - 1}}{x} \, dx.$$

6 Use the substitution $x = u^6$ to find

$$\int \frac{dx}{x^{1/2} + x^{1/3}}.$$

3.10 RATIONAL TRIGONOMETRIC FUNCTIONS

We first illustrate the problem with an example.

Example 3.10.1

$$\int \frac{\cos x \, dx}{2 + \sin x}.$$

Here the numerator is just the derivative of the denominator, so the substitution $u = 2 + \sin x$, $du = \cos x$ transforms the integral into

$$\int \frac{du}{u} = \ln |u| + c = \ln |2 + \sin x| + c.$$

This example was easy because of its special form, and we included it to emphasise that one should always be on the lookout for such methods. However, for more general rational functions of $\sin x$ and $\cos x$, the substitution $t = \tan(x/2)$ converts them into rational functions of t. The following identities enable us to do this:

$$\sin x = \frac{2t}{1 + t^2}; \ \cos x = \frac{1 - t^2}{1 + t^2}$$

Differentiating $t = \tan(x/2)$, we find

$$\frac{dt}{dx} = \frac{1}{2} \sec^2 \frac{x}{2} = \frac{1}{2} (1 + t^2)$$

so that

$$dx = \frac{2 \, dt}{1 + t^2}$$

Example 3.10.2

$$I = \int \frac{dx}{1 + \cos x}.$$

Putting $t = \tan(x/2)$, we obtain

$$I = \int \frac{1}{1 + (1 - t^2)/(1 + t^2)} \frac{2\,dt}{1 + t^2}$$

$$= \int \frac{2}{1 + t^2 + 1 - t^2}\,dt$$

$$= \int dt = \tan \frac{x}{2} + c$$

EXERCISE 3.10.1

Evaluate

$$\int \frac{dx}{\cos x + \sin x}.$$

(You will find Exercise 3.9.1.1(i) useful.)

3.11 IMPROPER INTEGRALS

A definite integral is called improper if either the range of integration is infinite or the integrand is infinite at one or more points of the range of integration. We study the two sorts of improper integral separately.

Infinite Range

Consider the integral $\int_a^b f(x)dx$. If the limit of this integral as $b \to \infty$ exists, we write

$$\lim_{b \to \infty} \int_a^b f(x)dx = \int_a^\infty f(x)dx$$

Example 3.11.1

Find $\int_0^\infty e^{-x}\,dx$ if it exists.

Figure 3.10 shows a sketch of e^{-x}; the integral is represented by the shaded area, and the question is as to whether this area is finite, so that the integral exists.

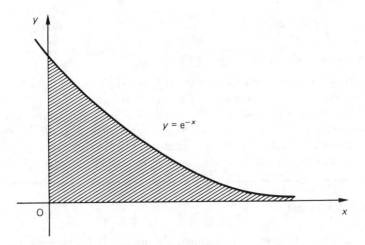

Fig. 3.10

Consider $\int_0^R e^{-x}\,dx = [-e^{-x}]_0^R = 1 - e^{-R}$. Letting $R \to \infty$, this tends to the value 1, the value of the integral, which clearly exists.

Provided that care is taken in handling limits, the usual methods of substitution and parts can be used for improper integrals. The first of the following examples is one that arises in the study of statistics.

═══════════════════════════════════════

Examples 3.11.2

1. Find $I_n = \int_0^\infty x^n e^{-x}\,dx$ for $n \geq 0$.

   We find a reduction formula for this integral. For $n > 0$, set $u = x^n$ and $dv = e^{-x}\,dx$, so that $du = nx^{n-1}\,dx$ and $v = -e^{-x}$. Then

   $$I_n = \lim_{R \to \infty} \int_0^R x^n e^{-x}\,dx$$

   $$= \lim_{R \to \infty} \left([-x^n e^{-x}]_0^R + n \int_0^R x^{n-1} e^{-x}\,dx\right)$$

   $$= \lim_{R \to \infty} (-R^n e^{-R}) + n \lim_{R \to \infty} \int_0^R x^{n-1} e^{-x}\,dx$$

   $$= n \int_0^\infty x^{n-1} e^{-x}\,dx$$

   $$= nI_{n-1}$$

where we have used

$$\frac{R^n}{e^R} = \frac{R^n}{1 + \cdots + R^{n+1}/(n+1)! + \cdots} \to 0 \text{ as } R \to \infty$$

Now $I_0 = 1$ from the last example, so $I_1 = 1$, $I_2 = 2!, \ldots, I_n = n!$

2. For what real values of p does $\int_0^\infty e^{px} \, dx$ exist?

If $p \neq 0$, then $\int_0^R e^{px} \, dx = (e^{pR} - 1)/p$. As $R \to \infty$, this tends to a finite limit $-1/p$ if $p < 0$, in which case the integral exists. If $p = 0$, $\int_0^R e^{px} \, dx = R \to \infty$ as $R \to \infty$, so the integral does not exist in this case.

The case of a lower limit of $-\infty$ can be treated similarly.

Infinite Integrand

Suppose that the function f appearing in the integral $\int_a^b f(x) \, dx$ goes to infinity at a point c in the interval (a, b). We write the integral in this case as $\int_a^c f(x) \, dx + \int_c^b f(x) \, dx$. Then $\int_a^b f(x) \, dx$ exists if each of these integrals exists. We need therefore only study the case where f has an infinity at the beginning or end of the range of integration. We concentrate on the former case, since the latter is treated in a similar way.

Let $I = \int_a^b f(x) \, dx$, where $f(x) \to \pm\infty$ as $x \to a$. Then what we actually mean is

$$I = \lim_{\varepsilon \to 0} \int_{a+\varepsilon}^b f(x) \, dx$$

if the limit exists.

Examples 3.11.3

1.
$$\int_\varepsilon^1 \frac{dx}{\sqrt{x}} = [2\sqrt{x}]_\varepsilon^1 = 2(1 - \sqrt{\varepsilon})$$

Letting $\varepsilon \to 0$, this has the limit 2. Thus,

$$\int_0^1 \frac{dx}{\sqrt{x}} = 2$$

2.
$$\int_0^{1-\varepsilon} (1-x)^{-2} \, dx = [(1-x)^{-1}]_0^{1-\varepsilon}$$

$$= \left(\frac{1}{\varepsilon} - 1\right)$$

and this clearly does not have a finite limit as $\varepsilon \to 0$, so the integral does not exist.

EXERCISES 3.11.1

1 For which real values of p does $\int_1^\infty x^p \, dx$ exist? Evaluate the integral for these values of p.

2 Evaluate the following integrals:

(i) $\displaystyle\int_0^\infty (x^2 + 1)^{-1} \, dx$; (ii) $\displaystyle\int_1^\infty x^3 e^{-x^4} \, dx$.

3 For which real values of p does $\int_0^1 x^p \, dx$ exist? Evaluate the integral for these values of p.

3.12 LIST OF STANDARD INTEGRALS

$$\int x^n \, dx = \begin{cases} \dfrac{x^{n+1}}{n+1} + c, & n \neq -1 \\ \ln|x| + c, & n = -1 \end{cases}$$

$$\int e^x \, dx = e^x + c$$

$$\int a^x \, dx = \frac{a^x}{\ln a} + c \qquad (a > 0, \, a \neq 1)$$

$$\int \sin x \, dx = -\cos x + c$$

$$\int \cos x \, dx = \sin x + c$$

$$\int \tan x \, dx = \ln|\sec x| + c$$

$$\int \sec x \, dx = \ln|\sec x + \tan x| + c$$

$$\int \operatorname{cosec} x \, dx = \ln|\operatorname{cosec} x - \cot x| + c$$

$$\int \cot x \, dx = \ln |\sin x| + c$$

$$\int \sec^2 x \, dx = \tan x + c$$

$$\int \sec x \tan x \, dx = \sec x + c$$

$$\int \sinh x \, dx = \cosh x + c$$

$$\int \cosh x \, dx = \sinh x + c$$

$$\int \tanh x \, dx = \ln \cosh x + c$$

$$\int \frac{dx}{\sqrt{a^2 - x^2}} = \sin^{-1} \frac{x}{a} + c \quad \left(a > 0, \, \left| \frac{x}{a} \right| < 1 \right)$$

$$\int \frac{dx}{\sqrt{x^2 + a^2}} = \sinh^{-1} \frac{x}{a} + c \quad (a > 0)$$

$$\int \frac{dx}{\sqrt{x^2 - a^2}} = \cosh^{-1} \frac{x}{a} + c \quad (x > a > 0)$$

$$\int \frac{dx}{x^2 + a^2} = \frac{1}{a} \tan^{-1} \frac{x}{a} + c \quad (a \neq 0)$$

MISCELLANEOUS EXERCISES 3

1 The *mean value* of a function f with respect to the variable x over an interval $[a, b]$ is given by $\dfrac{1}{b-a} \displaystyle\int_a^b f(x)dx$. Find the mean value of the displacement $x = a \sin \omega t$ of a simple pendulum (i) with respect to x over the interval $[0, a]$, (ii) with respect to t over the interval $\left[0, \dfrac{\pi}{2\omega} \right]$. Explain why these two answers differ.

2 The *root mean square* (rms for short) value of a function over an interval is the square root of the mean value of its square over the interval. Find the mean and rms values over a period π of the current in an electrical circuit given by $i = 2 + 3 \sin 2t$.

3 Show that the mean and rms values over a period $\dfrac{2\pi}{\omega}$ of the current given by $i = i_1 \sin \omega t + i_2 \sin 2\omega t$ are zero and $\sqrt{\dfrac{i_1^2 + i_2^2}{2}}$.

4 The work done by a gas expanding from a volume v_1 to a volume v_2 is $\int_{v_1}^{v_2} p \, dv$, where p is the pressure of the gas when it has volume v. Show that if $pv^\gamma = K$, where $\gamma \neq 1$ and K are constants, then the work done by the expanding gas is $\dfrac{K}{1 - \gamma}(v_2^{1-\gamma} - v_1^{1-\gamma})$.

5 The vapour pressure p of a liquid at temperature T K satisfies the equation $\dfrac{d}{dT} \ln p = \dfrac{LT^2}{R}$, where L is the latent heat of evaporation and R is the gas constant. Given that $p = p_1$ when $T = T_1$ and $p = p_2$ when $T = T_2$, integrate this equation from T_1 to T_2 assuming (a) L is constant, (b) $L = L_0 + aT$, where L_0 and a are constants.

6 In a particular second-order chemical reaction the reaction time t is related to the amount X of a product by the equation

$$kt = \int_0^X \frac{dx}{(1 - x)(3 - x)}$$

where k is the rate constant. Evaluate the integral and solve for X.

7 The electric force at a distance r along the axis from a circular disc of radius a carrying a uniform charge σ per unit area is $2\pi\sigma r \displaystyle\int_0^a \frac{x \, dx}{(x^2 + r^2)^{3/2}}$.

Evaluate this force and show that when a is very large compared with r it is approximately $2\pi\sigma$.

8 The electric potential V at a distance r from the centre of a sphere of radius a which carries a charge e uniformly distributed over its surface is given by

$$V = \frac{e}{2a} \int_{-a}^a \frac{dx}{\sqrt{a^2 - 2rx + r^2}}$$

Show that

$$V = \begin{cases} \dfrac{e}{a} & \text{if } r \leq a \\[2mm] \dfrac{e}{r} & \text{if } r > a \end{cases}$$

9 An infinitely long straight wire carrying an electric current i produces a magnetic force at a point P, distance r from the wire, of an amount

$$ir \int_{-\infty}^{\infty} \frac{dx}{(x^2 + r^2)^{3/2}}.$$ Show that this has the value $\frac{2i}{r}$.

10 The amount of a drug being eliminated from an animal at time t is $f(t)$. Assuming that it started taking the drug at time zero and that it eventually eliminated all of the drug, find the total amount of the drug administered to the animal when (a) $f(t) = Ae^{-kt}$, (b) $f(t) = At^2e^{-kt}$, where $A, k > 0$ are constants.

3.13 ANSWERS TO EXERCISES

Exercise 3.1.1

Let $x_i = \dfrac{i}{n}$, then $\displaystyle\int_0^1 x \, dx = \lim_{n \to \infty} \sum_{i=1}^{n} \frac{i}{n}\frac{1}{n} = \lim_{n \to \infty} \frac{1}{n^2} \sum_{i=1}^{n} i = \lim_{n \to \infty} \frac{n+1}{2n} =$

$\displaystyle\lim_{n \to \infty} \frac{1}{2}\left(1 + \frac{1}{n}\right) = \frac{1}{2}.$

Exercises 3.4.1

1 (i) 1, (ii) $[\ln|x|]_{-e^2}^{-e} = \ln \dfrac{e}{e^2} = \ln \dfrac{1}{e} = -1$, (iii) $\dfrac{3}{2}$, (iv) 0, (v) 2.

2 $\displaystyle\int_0^1 x \, dx = \left[\frac{x^2}{2}\right]_0^1 = \frac{1}{2}.$

3 For e^x, $\displaystyle\int \frac{x^n}{n!} = \frac{x^{n+1}}{(n+1)!}$, which is the $(n+1)$st term of the expansion of e^x. The constant of integration is found to be 1 by putting $x = 0$.

For $\cos x$, $\displaystyle\int (-1)^n \frac{x^{2n}}{(2n)!} = (-1)^n \frac{x^{2n+1}}{(2n+1)!}$, the nth term of the expansion of $\sin x$. The constant of integration is found to be zero by putting $x = 0$.

sin, cosh and sinh all follow the same pattern as cos.

Exercises 3.5.1

1 (i) With $u = x^2$, $du = 2x\,dx$ and the integral becomes

$$\int \cos u\,du = \sin u + C = \sin x^2 + C,$$

(ii) with $u = x^4$, $du = 4x^3\,dx$ and the integral becomes

$$\frac{1}{4}\int e^u\,du = \frac{1}{4}e^u + C = \frac{1}{4}e^{x^4} + C$$

(iii) with $u = \sin x$, $du = \cos x\,dx$, and the integral becomes

$$\int \frac{du}{u} = \ln|u| + C = \ln|\sin x| + C,$$

(iv) with $u = x^2 + 3x + 4$, $du = 2x + 3$ and the integral becomes

$$\int \frac{du}{u} = \ln|u| + C = \ln|x^2 + 3x + 4| + C,$$

(v) with $u = 1 - x^2$, $du = -2x\,dx$ and the integral becomes

$$-\frac{1}{2}\int \frac{du}{\sqrt{u}} = -\sqrt{u} + C = -\sqrt{1 - x^2} + C,$$

(vi) with $u = x^2 + 2x$, $du = 2(x + 1)$ and the integral becomes

$$\frac{1}{2}\int \frac{du}{u^3} = -\frac{1}{4u^2} + C = -\frac{1}{4(x^2 + 2x)^2} + C,$$

(vii) with $u = 3x + 2$, $du = 3$ and the integral becomes

$$\frac{1}{3}\int u^4\,du = \frac{1}{15}u^5 + C = \frac{1}{15}(3x + 2)^5 + C,$$

(viii) with $u = -5x$, $du = -5$ and the integral becomes

$$-\frac{1}{5}\int e^u\,du = -\frac{1}{5}e^u + C = -\frac{1}{5}e^{-5x} + C,$$

(ix) with $u = 7x$, $du = 7$ and the integral becomes

$$\frac{1}{7}\int \cos u\,du = \frac{1}{7}\sin u + C = \frac{1}{7}\sin 7x + C.$$

2 (i) $\displaystyle\int \sin 2x \sin 3x\,dx = \frac{1}{2}\int (\cos x - \cos 5x)dx = \frac{1}{2}\sin x - \frac{1}{10}\sin 5x + C,$

(ii) $\displaystyle\int \sin 2x \cos 3x\,dx = \frac{1}{2}\int (\sin 5x - \sin x)dx = -\frac{1}{10}\cos 5x + \frac{1}{2}\cos x + C.$

Exercises 3.6.1

1 (i) With $x = \tan u$, $dx = \sec^2 u\, du$, the integral becomes

$$\int (1 + \tan^2 u)^{-3/2} \sec^2 du = \int \cos u\, du$$

$$= \sin u + C = \frac{\tan u}{\sec u} + C = \frac{x}{\sqrt{1 + x^2}} + C,$$

(ii) with $x = 2 \cosh u$, $dx = 2 \sinh u\, du$ and the integral becomes

$$\int 4 \cosh^2 u (4 \sinh^2 u)^{-3/2}\, 2 \sinh u\, du$$

$$= \int \frac{\cosh^2 u}{\sinh^2 u}\, du = \int \coth^2 u\, du$$

$$= \int (1 + \operatorname{cosech}^2 u)du = u - \coth u + C$$

$$= u - \frac{\cosh u}{\sinh u} + C = \cosh^{-1} \frac{x}{2} - \frac{\dfrac{x}{2}}{\sqrt{\left(\dfrac{x}{2}\right)^2 - 1}} + C$$

$$= \cosh^{-1} \frac{x}{2} - \frac{x}{\sqrt{x^2 - 4}} + C,$$

(iii) with $x = \sin u$, $dx = \cos u\, du$ and the integral becomes

$$\int_0^{\pi/2} \sin^2 u \cos^2 u\, du = \frac{1}{4} \int_0^{\pi/2} \sin^2 2u\, du = \frac{1}{8} \int_0^{\pi/2} (1 - \cos 4u)du$$

$$= \frac{1}{8} \left[u - \frac{1}{4} \sin 4u \right]_0^{\pi/2} = \frac{\pi}{16}.$$

2 $g'(x) = \sec x \tan x + \sec^2 x = \sec x (\tan x + \sec x)$, so

$$\frac{g'(x)}{g(x)} = \sec x$$

and

$$\int \sec x\, dx = \int \frac{g'(x)}{g(x)}\, dx = \ln |g(x)| + C = \ln |\sec x + \tan x| + C.$$

Exercises 3.7.1

1 (i) With $u = x$, $dv = \sin x\, dx$ then $du = dx$, $v = -\cos x$, so

$$\int x \sin x\, dx = -x \cos x + \int \cos x\, dx = -x \cos x + \sin x + C,$$

(ii) with $u = x^2$, $dv = e^{3x}\, dx$ then

$$du = 2x\, dx, \quad v = \frac{1}{3}\, e^{3x},$$

so

$$\int x^2 e^{3x}\, dx = \frac{1}{3}\, x^2 e^{3x} - \frac{2}{3} \int x e^{3x}\, dx.$$

Taking

$$u_1 = x, \quad dv_1 = e^{3x}\, dx, \quad du_1 = dx, \quad v_1 = \frac{1}{3}\, e^{3x}$$

in this integral and integrating it by parts, we finally obtain the value

$$\frac{1}{3}\, x^2 e^{3x} - \frac{2}{9}\, x e^{3x} + \frac{2}{9} \int e^{3x}\, dx = \frac{1}{3}\, x^2 e^{3x} - \frac{2}{9}\, x e^{3x} + \frac{2}{27}\, e^{3x} + C,$$

(iii) with $u = \ln x$, $dv = x^4\, dx$ then

$$du = \frac{dx}{x}, \quad v = \frac{1}{5}\, x^5$$

and

$$\int x^4 \ln x\, dx = \frac{1}{5}\, x^5 \ln x - \frac{1}{5} \int x^5\, \frac{1}{x}\, dx$$

$$= \frac{1}{5}\, x^5 \ln x - \frac{1}{5} \int x^4\, dx = \frac{1}{5}\, x^5 \ln x - \frac{1}{25}\, x^5 + C.$$

(iv) Let

$$I = \int e^{2x} \cos 3x\, dx.$$

With

$$u = \cos 3x, \quad dv = e^{2x}\, dx, \quad du = -3 \sin 3x\, dx, \quad v = \frac{1}{2}e^{2x},$$

$$I = \frac{1}{2}\, e^{2x} \cos 3x + \frac{3}{2} \int e^{2x} \sin 3x\, dx$$

and integrating again by parts with

$$u = \sin 3x, \; dv = e^{2x}\,dx, \; du = 3\cos 3x\,dx, \; v = \frac{1}{2}e^{2x},$$

we find

$$I = \frac{1}{2}e^{2x}\cos 3x + \frac{3}{4}e^{2x}\sin 3x - \frac{9}{4}\int e^{2x}\cos 3x\,dx$$

$$= \frac{1}{2}e^{2x}\cos 3x + \frac{3}{4}e^{2x}\sin 3x - \frac{9}{4}I.$$

Finally, solving for I, we find

$$I = \frac{1}{13}e^{2x}(4\cos 3x + 3\sin 3x) + C.$$

(v)
$$I = \int \sec^3 x\,dx = \sec x \tan x - \int \sec x \tan^2 x\,dx$$

$$= \sec x \tan x + \int \sec x\,dx - I$$

$$= \sec x \tan x + \ln |\sec x + \tan x| - I,$$

so

$$I = \frac{1}{2}\sec x \tan x + \frac{1}{2}\ln |\sec x + \tan x| + C.$$

2 $I_n = \displaystyle\int \sin^{n-1} x \sin x\,dx = -\cos x \sin^{n-1} x + (n-1)\int \cos^2 x \sin^{n-2} x\,dx$

$$= -\cos x \sin^{n-1} x + (n-1)I_{n-2} - (n-1)I_n,$$

so $nI_n = (n-1)I_{n-2} - \cos x \sin^{n-1} x.$

$3I_3 = 2\displaystyle\int \sin x\,dx - \cos x \sin^2 x, \text{ so } I_3 = -\frac{1}{3}(2\cos x + \cos x \sin^2 x) + C$

$$I_4 = \frac{1}{8}(3x - 3\cos x \sin x - 2\cos x \sin^3 x) + C.$$

3 Using Exercise 3.7.1.2, we find $I_n = \dfrac{n-1}{n}I_{n-2}.$

Exercises 3.8.1

In most of the following solutions, the constant of integration C has been replaced by $\ln k$ and incorporated in the other terms.

1 (i) $\dfrac{1}{x+1} - \dfrac{1}{x+2}, \; \ln \left| \dfrac{k(x+1)}{x+2} \right|,$

(ii) $\dfrac{1}{2}\dfrac{1}{x}+\dfrac{1}{2}\dfrac{1}{x+2}, \dfrac{1}{2}\ln|kx(x+2)|,$

(iii) $\dfrac{1}{2a}\dfrac{1}{x-a}-\dfrac{1}{2a}\dfrac{1}{x+a}, \dfrac{1}{2a}\ln\left|\dfrac{k(x-a)}{x+a}\right|,$

(iv) $\dfrac{1}{2}\dfrac{1}{x-3}+\dfrac{1}{2}\dfrac{1}{x-1}, \dfrac{1}{2}\ln|k(x^2-4x+3)|,$

(v) $\dfrac{1}{x-2}+\dfrac{2}{(x-2)^2}, \ln|x-2|-\dfrac{2}{x-2}+C.$

2 (i) $\dfrac{1}{12}\dfrac{1}{x+1}-\dfrac{1}{3}\dfrac{1}{x-2}+\dfrac{1}{4}\dfrac{1}{x-3}, \dfrac{1}{12}\ln\left|\dfrac{k(x+1)(x-3)^3}{(x-2)^4}\right|$

(ii) $\dfrac{2}{x-1}+\dfrac{9}{(x-1)^2}+\dfrac{3}{(x-1)^3}, 2\ln|x-1|-\dfrac{9}{x-1}-\dfrac{3}{2}\dfrac{1}{(x-1)^2}+C.$

Exercises 3.9.1

1 (i) $\dfrac{1}{x^2-2x-1}=\dfrac{1}{\{x-1-\sqrt{2}\}\{x-1+\sqrt{2})}$

$$=\dfrac{1}{2\sqrt{2}}\dfrac{1}{x-1-\sqrt{2}}-\dfrac{1}{2\sqrt{2}}\dfrac{1}{x-1+\sqrt{2}},$$

so

$$\int\dfrac{dx}{x^2-2x-1}=\dfrac{1}{2\sqrt{2}}\ln\left|k\dfrac{x-1-\sqrt{2}}{x-1+\sqrt{2}}\right|$$

(ii) $x^2+2x+5=(x+1)^2+2^2$, so putting $x+1=2\tan\theta$ the integral becomes

$$\dfrac{1}{16}\int\dfrac{2\sec^2\theta\,d\theta}{\sec^4\theta}=\dfrac{1}{8}\int\cos^2\theta\,d\theta=\dfrac{1}{16}\int(1+\cos2\theta)d\theta$$

$$=\dfrac{1}{16}\left(\theta+\dfrac{1}{2}\sin2\theta\right)+C$$

$$=\dfrac{1}{16}\tan^{-1}\left(\dfrac{x+1}{2}\right)+\dfrac{1}{8}\dfrac{1}{8}\dfrac{x+1}{x^2+2x+5}+C.$$

(iii) $\int\dfrac{4x+7}{x^2+10x+29}dx=2\int\dfrac{(2x+10)dx}{x^2+10x+29}-13\int\dfrac{dx}{(x+5)^2+2^2}$

$$=2\ln|x^2+10x+29|-\dfrac{13}{2}\tan^{-1}\left(\dfrac{x+5}{2}\right)+C.$$

2 By long division, the integral becomes

$$\int \left(x^2 - x + 1 + \frac{3x + 17}{x^2 + 4x + 13} \right) dx$$

$$= \frac{1}{3} x^3 - \frac{1}{2} x^2 + x + \frac{3}{2} \int \frac{2x + 4}{x^2 + 4x + 13} \, dx + 11 \int \frac{dx}{(x + 2)^2 + 3^2}$$

$$= \frac{1}{3} x^3 - \frac{1}{2} x^2 + x + \frac{3}{2} \ln |x^2 + 4x + 13| + \frac{11}{3} \tan^{-1} \left(\frac{x + 2}{3} \right) + C.$$

3 Letting

$$\frac{2 - 6x}{x^4 - 1} = \frac{A}{x - 1} + \frac{B}{x + 1} + \frac{Cx + D}{x^2 + 1}$$

and following the usual procedure, we find

$$A = -1, \; B = -2, \; C = 3, \; D = -1.$$

Thus

$$\int \frac{2 - 6x}{x^4 - 1} dx = \int \left(-\frac{1}{x - 1} - \frac{2}{x + 1} + \frac{3x - 1}{x^2 + 1} \right) dx$$

$$= -\ln |x - 1| - 2 \ln |x + 1| + \frac{3}{2} \int \frac{2x \, dx}{x^2 + 1} - \int \frac{dx}{x^2 + 1}$$

$$= -\ln |x - 1| - 2 \ln |x + 1| + \frac{3}{2} \ln(x^2 + 1) - \tan^{-1} x + C.$$

4 With $u = e^x$, $du = e^x \, dx$, the integral becomes

$$\int \frac{u^2 - 1}{u + u^3} \frac{du}{u}.$$

Using the usual partial fraction procedure, we find

$$\int \frac{(u^2 - 1) du}{u^2(u^2 + 1)} = \int \left(-\frac{1}{u^2} + \frac{2}{u^2 + 1} \right) du$$

$$= \frac{1}{u} + 2 \tan^{-1} u + C$$

$$= e^{-x} + 2 \tan^{-1} e^x + C.$$

5 $x = 1 + u^2$, $dx = 2u\,du$, so

$$\int \frac{\sqrt{x-1}}{x}\,dx = 2 \int \frac{u^2\,du}{1+u^2} = 2 \int \left(1 - \frac{1}{1+u^2}\right) du$$

$$= 2u - 2\tan^{-1} u + C = 2\sqrt{x-1} - 2\tan^{-1}\sqrt{x-1} + C.$$

6 $dx = 6u^5\,du$, so the integral becomes, after cancelling u^2,

$$6\int \frac{u^3}{u+1}\,du = 6\int \left(u^2 - u + 1 - \frac{1}{u+1}\right) du$$

$$= 2u^3 - 3u^2 + 6u - 6\ln|u+1| + C$$

$$= 2\sqrt{x} - 3x^{1/3} + 6x^{1/6} - 6\ln|x^{1/6} + 1| + C.$$

Exercise 3.10.1

$$t = \tan \frac{x}{2}$$

transforms the integral into

$$\int \frac{2\,dt}{1-t^2+2t} = -2\int \frac{dt}{t^2-2t-1} = \frac{1}{\sqrt{2}}\ln\left|k\frac{t-1+\sqrt{2}}{t-1-\sqrt{2}}\right| + C,$$

from Exercise 3.9.1.1(i), or in terms of x,

$$= \frac{1}{\sqrt{2}}\ln\left|k\frac{\tan\dfrac{x}{2}-1+\sqrt{2}}{\tan\dfrac{x}{2}-1-\sqrt{2}}\right| + C.$$

Exercises 3.11.1

1 If $p = -1$,

$$\int_1^R \frac{dx}{x} = [\ln|x|]_1^R = \ln|R| \to \infty \text{ as } R \to \infty,$$

so the integral does not exist. If $p \neq -1$,

$$\int_1^R x^p\,dx = \left[\frac{x^{p+1}}{p+1}\right]_1^R = \frac{R^{p+1}}{p+1} - \frac{1}{p+1},$$

which has the finite limit

$$-\frac{1}{p+1} \quad \text{as} \quad R \to \infty \quad \text{if} \quad p < -1.$$

2 (i) $\displaystyle\int_0^R (x^2+1)^{-1}\,dx = [\tan^{-1} x]_0^R = \tan^{-1} R \to \frac{\pi}{2}$ as $R \to \infty$.

(ii) With $u = x^4$, $du = 4x^3\,dx$,

$$\int_1^R x^3 e^{-x^4}\,dx = \frac{1}{4}\int_1^{R^4} e^{-u}\,du = \frac{1}{4}[-e^{-u}]_1^{R^4} = \frac{1}{4}(-e^{-R^4} + e^{-1}) \to \frac{1}{4e}.$$

3 For $p = -1$,

$$\int_\varepsilon^1 \frac{dx}{x} = [\ln |x|]_\varepsilon^1 = \ln\left|\frac{1}{\varepsilon}\right|,$$

which has no limit as $\varepsilon \to 0$. For $p \neq -1$,

$$\int_\varepsilon^1 x^p\,dx = \left[\frac{x^{p+1}}{p+1}\right]_\varepsilon^1 = \frac{1}{p+1} - \frac{\varepsilon^{p+1}}{p+1}.$$

This has a limit of

$$\frac{1}{p+1} \quad \text{as} \quad \varepsilon \to 0 \quad \text{if} \quad p > -1.$$

Miscellaneous Exercises 3

1 (i) $\displaystyle\frac{1}{a}\int_0^a x\,dx = \frac{a}{2}.$

(ii) $\displaystyle\frac{2\omega}{\pi}\int_0^{\pi/2\omega} a \sin \omega t\,dt = \frac{2a}{\pi}[-\cos \omega t]_0^{\pi/2\omega} = \frac{2a}{\pi}.$

The difference is because x varies linearly with itself, but as a sine with t.

2 Mean $\displaystyle i = \frac{1}{\pi}\int_0^\pi (2 + 3\sin 2t)\,dt = \frac{1}{\pi}\left[2t - \frac{3}{2}\cos 2t\right]_0^\pi = 2.$

$$(\text{rms } i)^2 = \frac{1}{\pi} \int_0^\pi (2 + 3 \sin 2t)^2 \, dt$$

$$= \frac{1}{\pi} \int_0^\pi \left\{ 4 + 12 \sin 2t + \frac{9}{2}(1 - \cos 4t) \right\} dt$$

$$= \frac{1}{\pi} \left[4t - 6 \cos 2t + \frac{9}{2} t - \frac{9}{8} \sin 4t \right]_0^\pi$$

$$= \frac{17}{2}.$$

Thus rms $i = \sqrt{\frac{17}{2}}$.

3 $$\text{mean } i = \frac{\omega}{2\pi} \int_0^{2\pi/\omega} (i_1 \sin \omega t + i_2 \sin 2\omega t) dt$$

$$= \frac{1}{2\pi} \left[-i_1 \cos \omega t - \frac{1}{2} i_2 \cos 2\omega t \right]_0^{2\pi/\omega}$$

$$= 0$$

$$(\text{rms } i)^2 = \frac{\omega}{2\pi} \int_0^{2\pi/\omega} (i_1^2 \sin^2 \omega t + 2i_1 i_2 \sin \omega t \sin 2\omega t + i_2^2 \sin^2 2\omega t) dt$$

$$= \frac{\omega}{2\pi} \int_0^{2\pi/\omega} \left\{ \frac{1}{2} i_1^2 (1 - \cos 2\omega t) + i_1 i_2 (\cos \omega t - \cos 3\omega t) \right.$$

$$\left. + \frac{1}{2} i_2^2 (1 - \cos 4\omega t) \right\} dt$$

$$= \frac{\omega}{2\pi} \left[\frac{1}{2} i_1^2 \left(t - \frac{1}{2\omega} \sin 2\omega t \right) + \frac{i_1 i_2}{\omega} \left(\sin \omega t - \frac{1}{3} \sin 3\omega t \right) \right.$$

$$\left. + i_2^2 \left(t - \frac{1}{4\omega} \sin 4\omega t \right) \right]_0^{2\pi/\omega}$$

$$= \frac{1}{2} (i_1^2 + i_2^2).$$

4 Work $= \int_{v_1}^{v_2} K v^{-\gamma} \, dv = \frac{K}{1-\gamma} (v_2^{1-\gamma} - v_1^{1-\gamma}).$

5 (a) $\ln p = \dfrac{L}{R} \displaystyle\int T^2 \, dT = \dfrac{LT^3}{3R} + C$. Putting in this $p = p_1$ when $T = \acute{T}_1$,

$p = p_2$ when $T = T_2$ and subtracting, we find $\ln\left(\dfrac{p_2}{p_1}\right) = \dfrac{L}{3R}(T_2^3 - T_1^3)$,

so $p_2 = p_1 e^{(L/3R)(T_2^3 - T_1^3)}$.

(b) $\ln\left(\dfrac{p_2}{p_1}\right) = \dfrac{1}{R}\displaystyle\int_{T_1}^{T_2} (L_0 + aT)T^2 \, dT = \dfrac{1}{R}\left[\dfrac{1}{3}L_0 T^3 + \dfrac{1}{4}aT^4\right]_{T_1}^{T_2}$, so

$p_2 = p_1 e^{(1/R)\{1/3 L_0(T_2^3 - T_1^3) + (1/4)a(T_2^4 - T_1^4)\}}$.

6 $kt = \dfrac{1}{2}\displaystyle\int_0^X \left(\dfrac{1}{1-x} - \dfrac{1}{3-x}\right) dx = \dfrac{1}{2}\left[\ln\left|\dfrac{3-x}{1-x}\right|\right]_0^X = \dfrac{1}{2}\ln\left|\dfrac{1 - \frac{1}{3}X}{1 - X}\right|$.

If $X < 1$ or $X > 3$, $\dfrac{1 - \frac{1}{3}X}{1 - X} = e^{2kt} \Rightarrow X = \dfrac{e^{2kt} - 1}{e^{2kt} - \frac{1}{3}}$. If $1 < X < 3$, the sign of

e^{2kt} must be changed, so $X = \dfrac{e^{2kt} + 1}{e^{2kt} + \frac{1}{3}}$.

7 $\text{Force} = 2\pi\sigma r \displaystyle\int_0^a \dfrac{x \, dx}{(x^2 + r^2)^{3/2}} = 2\pi\sigma r\left[-\dfrac{1}{\sqrt{x^2 + r^2}}\right]_0^a$

$= 2\pi\sigma\left(1 - \dfrac{1}{\sqrt{1 + (a^2/r^2)}}\right) \to 2\pi\sigma \quad \text{as} \quad a \to \infty.$

8 $V = \dfrac{e}{2a}\displaystyle\int_{-a}^{a} \dfrac{dx}{\sqrt{a^2 + r^2 - 2rx}}$

$= \dfrac{e}{2a}\left[-\dfrac{1}{r}\sqrt{a^2 + r^2 - 2rx}\right]_{-a}^{a}$

$= \dfrac{e}{2ar}\{-\sqrt{(a-r)^2} + \sqrt{(a+r)^2}\}$

$= \begin{cases} \dfrac{e}{a} & r \le a \\[2mm] \dfrac{e}{r} & r > a \end{cases}$

since $\sqrt{(a-r)^2}$ equals $a - r$ if $a > r$, but $r - a$ if $a < r$.

94

9 Force $= ir \displaystyle\int_{-R}^{R} \dfrac{dx}{(x^2+r^2)^{3/2}}$. The substitution $x = r \tan\theta$, $dx = r\sec^2\theta\, d\theta$

transforms this into

$$\frac{i}{r} \int_{-\tan^{-1}R/r}^{\tan^{-1}R/r} \cos\theta\, d\theta$$

$$= \frac{i}{r}[\sin\theta]_{-\tan^{-1}R/r}^{\tan^{-1}R/r}$$

$$= \frac{i}{r}\left\{ \frac{R}{\sqrt{r^2+R^2}} - \frac{-R}{\sqrt{r^2+R^2}} \right\}$$

$$= \frac{2i}{r\sqrt{1+r^2/R^2}} \to \frac{2i}{r} \quad \text{as} \quad R \to \infty.$$

10 Total drug eliminated $= \displaystyle\int_{0}^{\infty} f(t)\, dt$.

(a) $\displaystyle\int_{0}^{\infty} f(t)dt = A \int_{0}^{\infty} e^{-kt}\, dt = \frac{A}{k}[-e^{-kt}]_{0}^{\infty} = \frac{A}{k}.$

(b) $\displaystyle\int_{0}^{\infty} f(t)dt = A \int_{0}^{\infty} t^2 e^{-kt}\, dt$

$$= A\left\{ \left[-\frac{t^2 e^{-kt}}{k} \right]_{0}^{\infty} + \frac{2}{k} \int_{0}^{\infty} t e^{-kt}\, dt \right\}$$

$$= A\left\{ \left[-\frac{2}{k^2} t e^{-kt} \right]_{0}^{\infty} + \frac{2}{k^2} \int_{0}^{\infty} e^{-kt}\, dt \right\}$$

$$= \frac{2A}{k^3}.$$

[Note that it can be proved that $\lim_{t\to\infty} t^n e^{-kt} = 0$ for any fixed n, $k > 0$.]

4 LINEAR EQUATIONS, DETERMINANTS AND MATRICES

4.1 INTRODUCTION

The material of this chapter provides some of the most useful mathematical tools available for scientists, engineers and, indeed, social scientists. We start with a series of examples to illustrate the different types of solution that two simultaneous equations can have.

Examples 4.1.1

1a.
$$2x - 3y = 7 \tag{1}$$
$$3x + 5y = 1 \tag{2}$$

In order to eliminate x, we subtract $\frac{3}{2}$ times Equation (1) from Equation (2); we shall describe this operation in abbreviated form as

$$(2) - \tfrac{3}{2} \times (1)$$

This operation gives $\frac{19}{2}y = -\frac{19}{2}$, and hence $y = -1$; putting this back into Equation (1), we find $x = 2$.

1b.
$$2x - 3y = 0 \tag{1}$$
$$3x + 5y = 0 \tag{2}$$

These equations have the same left-hand sides as those in Example 1a; we describe them as the *homogeneous* set of equations corresponding to the set in Example 1a. The operation $(2) - \frac{3}{2} \times (1)$ now yields $\frac{19}{2}y = 0$ and hence $y = 0$. Equation (1) then gives $x = 0$ also. We describe this as the *trivial solution*.

2a.
$$2x - 3y = 7 \tag{1}$$
$$4x - 6y = 1 \tag{2}$$

The operation $(2) - 2 \times (1)$ gives $0 = -13$! Clearly, no solution is possible; indeed, multiplying Equation (1) through by 2 gives

$$4x - 6y = 14$$

which actually contradicts Equation (2). We say in this case that the equations are *inconsistent*.

2b. The homogeneous set corresponding to the equations in Example 2a are

$$2x - 3y = 0 \qquad (1)$$

$$4x - 6y = 0 \qquad (2)$$

and it is clear that Equation (2) is Equation (1) multiplied by 2, so that values of x and y that satisfy Equation (1) automatically satisfy Equation (2); effectively, there is only one equation to satisfy. In order to describe the solution, we let y be an arbitrary number, λ, say; then, from either equation, x must take the value $3\lambda/2$. We can write the solution as $(x, y) = (3\lambda/2, \lambda)$ for any value of λ. This represents an infinity of solutions.

2c.
$$2x - 3y = 7 \qquad (1)$$

$$4x - 6y = 14 \qquad (2)$$

This set comes from Example 2a, but the right-hand side of the second equation has been modified to make the two equations consistent. We describe the solution in exactly the same way as for the homogeneous case: let $y = \lambda$; then, from Equation (1), we find $x = 7/2 + 3\lambda/2$. The infinity of solutions is thus given by $(7/2 + 3\lambda/2, \lambda)$ for any λ.

Let us analyse the case of two simultaneous linear equations; suppose they are

$$ax + by = c \qquad (1)$$

$$dx + ey = f \qquad (2)$$

If $a \neq 0$, the operation $(2) - \dfrac{d}{a} \times (1)$ yields

$$\left(e - \frac{bd}{a} \right) y = f - \frac{cd}{a}$$

which gives, after multiplying through by a,

$$(ae - bd)y = (af - cd) \qquad (3)$$

and this equation also holds if $a = 0$. If $ae - bd \neq 0$, we find that

$$y = \frac{af - cd}{ae - bd} \qquad (4)$$

97

Putting this value for y into Equation (1), we find that

$$x = \frac{ce - bf}{ae - bd} \tag{5}$$

It is useful at this stage to write the denominators of (4) and (5) as

$$ae - bd = \begin{vmatrix} a & b \\ d & e \end{vmatrix}$$

the *determinant of coefficients*. This equation actually defines the *second-order determinant*,

$$\begin{vmatrix} a & b \\ d & e \end{vmatrix}$$

Its value is found by cross-multiplying the elements, a by e, and b by d, and subtracting. The numbers a, b, c and d are called the *elements* of the determinant.

We can also write the numerator of the right-hand side of Equation (4)

$$af - cd = \begin{vmatrix} a & c \\ d & f \end{vmatrix}$$

and Equation (4) now becomes

$$y = \frac{\begin{vmatrix} a & c \\ d & f \end{vmatrix}}{\begin{vmatrix} a & b \\ d & e \end{vmatrix}} \qquad \text{if} \quad \begin{vmatrix} a & b \\ d & e \end{vmatrix} \neq 0$$

Similarly, Equation (5) becomes

$$x = \frac{\begin{vmatrix} c & b \\ f & e \end{vmatrix}}{\begin{vmatrix} a & b \\ d & e \end{vmatrix}} \qquad \text{if} \quad \begin{vmatrix} a & b \\ d & e \end{vmatrix} \neq 0$$

For a unique solution, then, we must have that the determinant of coefficients is non-zero. In the corresponding homogeneous case, $c = f = 0$, and it is easy to see from the definition that

$$\begin{vmatrix} c & b \\ f & e \end{vmatrix} = \begin{vmatrix} a & c \\ d & f \end{vmatrix} = 0$$

giving $x = y = 0$.

If, however, the determinant of coefficients vanishes, we still have

Equation (3), which in determinant form is

$$\begin{vmatrix} a & b \\ d & e \end{vmatrix} y = \begin{vmatrix} a & c \\ d & f \end{vmatrix}$$

and a similar equation satisfied by x,

$$\begin{vmatrix} a & b \\ d & e \end{vmatrix} x = \begin{vmatrix} c & b \\ f & e \end{vmatrix}$$

Since the coefficients of y and x in these equations are both zero, the only way they can be satisfied is by the right-hand sides also being zero. This is certainly true in the homogeneous case, $c = f = 0$, which leads to the infinity of solutions. When c and f are not zero, we must have for a solution

$$\begin{vmatrix} a & b \\ d & e \end{vmatrix} = \begin{vmatrix} c & b \\ f & e \end{vmatrix} = \begin{vmatrix} a & c \\ d & f \end{vmatrix} = 0$$

which may be rearranged into

$$ae - bd = ce - bf = af - cd = 0$$

These equations are also equivalent to

$$\frac{a}{d} = \frac{b}{e} = \frac{c}{f}$$

which shows that the equations must be proportional for a solution.

We shall postpone investigation of the problem for three or more equations until we have introduced higher-order determinants and some of their properties.

EXERCISES 4.1.1

1 Solve the following sets of simultaneous equations:

(i) $x - y = 1$
$x + y = 3$;

(ii) $x + 2y = 0$
$3x + 6y = 0$;

(iii) $x + 2y = 1$
$3x + 6y = 2$;

(iv) $x + 2y = 1$
$3x + 6y = 3$.

2 Evaluate the determinants

(i) $\begin{vmatrix} 1 & -1 \\ 0 & -2 \end{vmatrix}$; (ii) $\begin{vmatrix} 3 & 5 \\ -1 & 2 \end{vmatrix}$; (iii) $\begin{vmatrix} 4 & 1 \\ -1 & -2 \end{vmatrix}$; (iv) $\begin{vmatrix} \cos\theta & \sin\theta \\ -\sin\theta & \cos\theta \end{vmatrix}$.

4.2 DETERMINANTS

We define a third-order determinant by its expansion in second order determinants:

$$\begin{vmatrix} a & b & c \\ d & e & f \\ g & h & i \end{vmatrix} = a \begin{vmatrix} e & f \\ h & i \end{vmatrix} - b \begin{vmatrix} d & f \\ g & i \end{vmatrix} + c \begin{vmatrix} d & e \\ g & h \end{vmatrix}$$

This is called the *expansion about the first row*. We note that the signs alternate in this expansion and that each element is multiplied by the second-order determinant obtained by excluding the row and column containing that element. For example, the element b is in the first row and second column and therefore multiplies the determinant

$$\begin{vmatrix} d & f \\ g & i \end{vmatrix} \tag{1}$$

which is obtained by excluding the first row and second column of the larger determinant. This is called the *minor* of b. When the appropriate sign is attached, we call the signed minor

$$- \begin{vmatrix} d & f \\ g & i \end{vmatrix}$$

the *cofactor* of b. The expansion can thus be described as the sum of the products of the elements of the first row by the corresponding cofactors.

This serves to define a third-order determinant, but it is easy to check (by multiplying out all the terms) that it can also be expanded about any other row or about any column. We must, however, give a rule for determining the sign of each cofactor; the cofactor of the element in the ith row and jth column is found by multiplying the appropriate minor by $(-1)^{i+j}$. In the expansion about the first row above, for example, b is in the first row and second column, so the cofactor of b is

$$(-1)^{1+2} \begin{vmatrix} d & f \\ g & i \end{vmatrix} = - \begin{vmatrix} d & f \\ g & i \end{vmatrix}$$

The expansion about the second column is

$$\begin{vmatrix} a & b & c \\ d & e & f \\ g & h & i \end{vmatrix} = -b \begin{vmatrix} d & f \\ g & i \end{vmatrix} + e \begin{vmatrix} a & c \\ g & i \end{vmatrix} - h \begin{vmatrix} a & c \\ d & f \end{vmatrix}$$

If we draw an array of these signs, they form a chessboard-like pattern, shown below for the order 5 case:

$$
\begin{array}{ccccc}
+ & - & + & - & + \\
- & + & - & + & - \\
+ & - & + & - & + \\
- & + & - & + & - \\
+ & - & + & - & +
\end{array}
$$

The following properties are easily deduced from the definition:

(1) Multiplying every element of a row (or column) by a number k multiplies the value of the determinant by k. For example, expanding about the second row,

$$
\begin{vmatrix} a & b & c \\ kd & ke & kf \\ g & h & i \end{vmatrix} = -kd \begin{vmatrix} b & c \\ h & i \end{vmatrix} + ke \begin{vmatrix} a & c \\ g & i \end{vmatrix} - kf \begin{vmatrix} a & b \\ g & h \end{vmatrix}
$$

$$
= k \begin{vmatrix} a & b & c \\ d & e & f \\ g & h & i \end{vmatrix}
$$

(2) If two rows (or two columns) are identical, then the determinant has value zero. For example, expanding about the first column,

$$
\begin{vmatrix} a & b & b \\ d & e & e \\ g & h & h \end{vmatrix} = a \begin{vmatrix} e & e \\ h & h \end{vmatrix} - d \begin{vmatrix} b & b \\ h & h \end{vmatrix} + g \begin{vmatrix} b & b \\ e & e \end{vmatrix}
$$

$$
= 0
$$

(3) If one row (or column) is a multiple of another row (or column) the determinant has the value zero. This is just a combination of properties 1 and 2 above.

(4) Adding a multiple of one row (or column) to another row (or column) does not alter the value of the determinant. For example, expanding

101

about the first row,

$$\begin{vmatrix} a+kd & b+ke & c+kf \\ d & e & f \\ g & h & i \end{vmatrix} = (a+kd)\begin{vmatrix} e & f \\ h & i \end{vmatrix} - (b+ke)\begin{vmatrix} d & f \\ g & i \end{vmatrix}$$

$$+ (c+kf)\begin{vmatrix} d & e \\ g & h \end{vmatrix}$$

$$= a\begin{vmatrix} e & f \\ h & i \end{vmatrix} - b\begin{vmatrix} d & f \\ g & i \end{vmatrix} + c\begin{vmatrix} d & e \\ g & h \end{vmatrix}$$

$$+ kd\begin{vmatrix} e & f \\ h & i \end{vmatrix} - ke\begin{vmatrix} d & f \\ g & i \end{vmatrix} + kf\begin{vmatrix} d & e \\ g & h \end{vmatrix}$$

$$= \begin{vmatrix} a & b & c \\ d & e & f \\ g & h & i \end{vmatrix} + k\begin{vmatrix} d & e & f \\ d & e & f \\ g & h & i \end{vmatrix}$$

and the second of these determinants has the value zero, since it has two equal rows.

Examples 4.2.1

1. Evaluate the determinant

$$\begin{vmatrix} 1 & 3 & 4 \\ 2 & 2 & 5 \\ 3 & 4 & 1 \end{vmatrix}$$

We could expand this directly to obtain the value, but we prefer to use determinant properties to simplify it first. Thus, subtracting twice the first row from the second and three times the first row from the last, we obtain, after expanding about the first column,

$$\begin{vmatrix} 1 & 3 & 4 \\ 0 & -4 & -3 \\ 0 & -5 & -11 \end{vmatrix} = \begin{vmatrix} -4 & -3 \\ -5 & -11 \end{vmatrix} = 29$$

The idea is to obtain a row (or column) with only one non-zero element.

2. Evaluate

$$\begin{vmatrix} 1 & 1 & 1 \\ a & b & c \\ a^2 & b^2 & c^2 \end{vmatrix}$$

Subtracting the first column from the second and third columns, we find

$$
\begin{vmatrix} 1 & 1 & 1 \\ a & b & c \\ a^2 & b^2 & c^2 \end{vmatrix} = \begin{vmatrix} 1 & 0 & 0 \\ a & b-a & c-a \\ a^2 & b^2-a^2 & c^2-a^2 \end{vmatrix}
$$

$$
= \begin{vmatrix} b-a & c-a \\ b^2-a^2 & c^2-a^2 \end{vmatrix}
$$

$$
= (b-a)(c-a)\begin{vmatrix} 1 & 1 \\ b+a & c+a \end{vmatrix}
$$

$$
= (b-a)(c-a)(c-b)
$$

where we expanded about the first row to obtain the second line and took factors $b-a$ and $c-a$ out of the first and second columns, respectively, in the third line.

Determinants of order higher than 3 are defined in the same way as the sum of the products of elements in a given row (or column) by their cofactors. The rule for signs is as already stated for the order 3 case, and the minor for each element is the determinant obtained by excluding the row and column containing that element.

EXERCISES 4.2.1

1 Evaluate the determinants

(i) $\begin{vmatrix} 7 & 11 & 4 \\ 13 & 15 & 10 \\ 3 & 9 & 6 \end{vmatrix}$; (ii) $\begin{vmatrix} 1 & 2 & 3 \\ 2 & 3 & 4 \\ 3 & 4 & 5 \end{vmatrix}$.

2 Find values of x which satisfy the equation

$$
\begin{vmatrix} -x & 1 & 0 \\ 1 & -x & 1 \\ 0 & 1 & -x \end{vmatrix} = 0
$$

4.3 THREE OR MORE SIMULTANEOUS EQUATIONS

We start by solving an example containing a number of parameters, which can be set to illustrate several different cases.

Example 4.3.1

$$x + 5y + 3z = a \qquad (1)$$

$$x + 2y + kz = b \qquad (2)$$

$$5x + y - kz = c \qquad (3)$$

where k, a, b and c are parameters.

There are many ways of eliminating the variables; we choose to do it in a particular order—first, because this enables us to link in determinants to the problem, and second, because it corresponds with an algorithm described later for the practical solution of equations. We start by eliminating x from the second two equations by taking suitable multiples of the first equation from them:

$$x + 5y + 3z = a \qquad (1)$$

$$-3y + (k - 3)z = b - a \qquad (2)' = (2) - (1)$$

$$-24y - (k + 15)z = c - 5a \qquad (3)' = (3) - 5 \times (1)$$

Notice that we have repeated Equation (1) for convenient reference, even though we have not altered it. We now eliminate y from Equation (3)' with the help of Equation (2)':

$$x + 5y + 3z = a \qquad (1)$$

$$-3y + (k - 3)z = b - a \qquad (2)'$$

$$9(1 - k)z = 3a - 8b + c \qquad (3)'' = (3)' - 8 \times (2)'$$

If $k \neq 1$, we find the value of z from Equation (3)''; putting this into Equation (2)' gives the value of y, and putting both of these values into Equation (1) gives the value of x. In this case the solution is unique.

We obtain the solution of the corresponding homogeneous case by putting $a = b = c = 0$, giving $x = y = z = 0$, that is, the trivial solution.

Now consider the case when $k = 1$; from Equation (3)'', there can be no solution if $3a - 8b + c \neq 0$, when the equations are inconsistent. If this quantity is zero, Equation (3)'' becomes $0 = 0$, so that we effectively have only two equations for x, y and z to satisfy. We obtain an infinite number of answers from Equations (1) and (2)' by letting z take an arbitrary value, λ, say, just as in the case of two equations. We easily obtain, from Equations (2)' and (1), with $k = 1$,

$$y = \frac{a - b}{3} - \frac{2}{3}\lambda$$

$$x = a - 5y - 3z = \frac{5b - 2a}{3} + \frac{\lambda}{3}$$

The solution is thus

$$\left(\frac{5b-2a}{3} + \frac{\lambda}{3}, \frac{a-b}{3} - \frac{2}{3}\lambda, \lambda \right) \quad \text{for any value of } \lambda$$

We now relate these cases to the determinant of coefficients, which is

$$\begin{vmatrix} 1 & 5 & 3 \\ 1 & 2 & k \\ 5 & 1 & -k \end{vmatrix}$$

The operations used in eliminating x and y from Equations (2) and (3) change the determinant of coefficients, but do not alter its value, since they simply subtract multiples of rows from other rows (see property 4 of determinants). Thus, the determinant of coefficients has the value of the final determinant,

$$\begin{vmatrix} 1 & 5 & 3 \\ 0 & -3 & k-3 \\ 0 & 0 & 9(1-k) \end{vmatrix} = \begin{vmatrix} -3 & k-3 \\ 0 & 9(1-k) \end{vmatrix}$$

$$= 27(k-1)$$

We found earlier that the equations had a unique solution if $k \neq 1$, and this corresponds to the non-vanishing of the determinant of coefficients, just as in the case of two equations. Although this is only an example and therefore does not prove anything, it does nevertheless suggest the following general result, which can be proved with more sophisticated methods.

A set of n simultaneous linear equations in n variables has a unique solution if the determinant of its coefficients is non-zero (the corresponding homogeneous set has only the trivial solution). If this determinant vanishes, then the equations have an infinity of solutions if the equations are consistent (which is always true in the homogeneous case) or no solutions if the equations are inconsistent.

EXERCISES 4.3.1

1 Solve the following systems of linear equations:

$$2x + 3y - z = -1 \qquad\qquad x + 2y - z = 1$$
(i) $\quad -x - 4y + 5z = 3 \quad ; \qquad$ (ii) $\quad 3x - 5y + 2z = 6 \quad ;$
$$x - 2y - 3z = 3 \qquad\qquad -x + 9y - 4z = -4$$

$$x + 3y + z = 0 \qquad\qquad x + y + z = 1$$

(iii) $2x - y - z = 1;$ (iv) $2x + 2y + 2z = 2.$

$$x - 4y - 2z = 0 \qquad\qquad 5x + 5y + 5z = 5$$

(*Hint* for (iv): Let y and z each be arbitrary numbers.)

2 Give solutions where possible of the system of linear equations

$$x + y + z = 3$$
$$x + 2y + 2z = 5$$
$$x + ay + bz = 3$$

for each of the following pairs of values of a and b:

(i) $a = b = 1;$ (ii) $a = 1, b \neq 1;$ (iii) $a \neq 1, b = 1;$
(iv) $a = b \neq 1;$ (v) $a \neq 1, b \neq 1, a \neq b.$

4.4 MATRICES

We now move a stage further in the solution of linear equations by introducing matrix notation. We start by representing the coefficients in the equations

$$ax + by = c$$
$$dx + ey = f$$

by the array

$$\begin{bmatrix} a & b \\ d & e \end{bmatrix}$$

Similarly, for the equations in Example 4.3.1, we represent the coefficients by the array

$$\begin{bmatrix} 1 & 5 & 3 \\ 1 & 2 & k \\ 5 & 1 & -k \end{bmatrix} = B, \text{ say}$$

We call each of these arrays the *matrix* of coefficients of the corresponding set of equations. The numbers appearing in the matrix are called its elements. Note that, although they look very similar to the determinants of coefficients, they are in fact very different objects. For a determinant is just a way of writing a single number, that is, its value, while this new object is an array of numbers, which cannot be represented by a single number. The different ways of writing the two should be carefully noted.

Note also that we have introduced the single capital letter, B, as a shorthand notation for the second matrix. This idea will turn out to be very useful in later manipulations.

We do not have to have a set of equations to define a matrix: *any rectangular array of numbers constitutes a matrix.*

A matrix with a single row is called a *row vector*, and a matrix with a single column is called a *column vector*. We usually call the numbers appearing in a vector *components* rather than elements, as in a matrix. We use a small letter in **bold** type as an abbreviation. For example, $\mathbf{a} = [1 \quad 2 \quad 3]$ is a row vector, while

$$\mathbf{s} = \begin{bmatrix} 1 \\ 2 \\ 3 \end{bmatrix}$$

is a column vector.

A matrix with m rows and n columns is called an $m \times n$ matrix. If $m = n$, the matrix is said to be square.

Now let

$$B = \begin{bmatrix} 1 & 5 & 3 \\ 1 & 2 & k \\ 5 & 1 & -k \end{bmatrix}, \quad \mathbf{r} = \begin{bmatrix} x \\ y \\ z \end{bmatrix} \quad \text{and} \quad \mathbf{s} = \begin{bmatrix} a \\ b \\ c \end{bmatrix}$$

Then we abbreviate the set of equations

$$\left. \begin{array}{l} x + 5y + 3z = a \\ x + 2y + kz = b \\ 5x + y - kz = c \end{array} \right\} \tag{1}$$

by the equation

$$B\mathbf{r} = \mathbf{s} \tag{2}$$

We can regard this as a shorthand way of writing the full equations, but it is useful to interpret it in a more specific way. We want the single matrix-vector Equation (2) to represent the three separate equations (1). Since the right-hand side $\mathbf{s}$ has three components, we should like the left-hand side also to have three components, so that we can obtain the three equations by equating corresponding components on each side of Equation (2). For this to happen, we must have

$$B\mathbf{r} = \begin{bmatrix} x + 5y + 3z \\ x + 2y + kz \\ 5x + y - kz \end{bmatrix}$$

We can achieve this by defining $B\mathbf{r}$ as the product of B and $\mathbf{r}$, with multiplication defined in the following way:

The ith component of $B\mathbf{r}$ is obtained by multiplying the elements of the ith row of B by corresponding components of $\mathbf{r}$ and adding.

107

We now tidy up and generalise this definition. Let A be the $m \times n$ matrix in which the element in the ith row and jth column is a_{ij} for $i = 1, 2, \ldots, m$ and $j = 1, 2, \ldots, n$. We write $A = [a_{ij}]$. Letting also

$$\mathbf{r} = \begin{bmatrix} x_1 \\ x_2 \\ \vdots \\ x_n \end{bmatrix} \qquad \mathbf{s} = \begin{bmatrix} s_1 \\ s_2 \\ \vdots \\ s_n \end{bmatrix}$$

we define the ith component of the product $A\mathbf{r}$ as

$$a_{i1}x_1 + a_{i2}x_2 + \cdots + a_{in}x_n,$$

which we write more concisely as $\sum_{j=1}^{n} a_{ij}x_j$. The component equations of $A\mathbf{r} = \mathbf{s}$ are thus

$$\sum_{j=1}^{n} a_{ij}x_j = s_i, \qquad i = 1, 2, \ldots, m$$

Example 4.4.1

Write the following equations in matrix form:

$$r - 2s + 3t = 4$$
$$5r + 6s - 7t = 8$$

The matrix form is

$$\begin{bmatrix} 1 & -2 & 3 \\ 5 & 6 & -7 \end{bmatrix} \begin{bmatrix} r \\ s \\ t \end{bmatrix} = \begin{bmatrix} 4 \\ 8 \end{bmatrix}$$

Addition of Matrices

Suppose we are given some equations in the form

$$\left. \begin{aligned} 2x + 3y + 4z &= 5 - (x - 2y + 3z) \\ 3x + 4y + 5z &= 6 - (2x - 3y - 4z) \\ 4x + 5y + 6z &= 7 - (3x + 4y + 5z) \end{aligned} \right\} \tag{1}$$

which we can write in matrix form as

$$A\mathbf{r} = \mathbf{b} - B\mathbf{r} \tag{2}$$

where

$$A = \begin{bmatrix} 2 & 3 & 4 \\ 3 & 4 & 5 \\ 4 & 5 & 6 \end{bmatrix}, \quad B = \begin{bmatrix} 1 & -2 & 3 \\ 2 & -3 & -4 \\ 3 & 4 & 5 \end{bmatrix}, \quad \mathbf{r} = \begin{bmatrix} x \\ y \\ z \end{bmatrix} \quad \text{and} \quad \mathbf{b} = \begin{bmatrix} 5 \\ 6 \\ 7 \end{bmatrix}$$

Clearly, we should simplify the original set of equations (1) by collecting like terms on the left-hand side, to obtain

$$3x + y + 7z = 5$$

$$5x + y + z = 6$$

$$x + 9y + 11z = 7$$

which we write as

$$C\mathbf{r} = \mathbf{b} \tag{3}$$

where C is the matrix

$$\begin{bmatrix} 3 & 1 & 7 \\ 5 & 1 & 1 \\ 1 & 9 & 11 \end{bmatrix}$$

By adding the vector $B\mathbf{r}$ to both sides of Equation (2), we obtain

$$A\mathbf{r} + B\mathbf{r} = \mathbf{b}$$

which we should like to write in the form

$$(A + B)\mathbf{r} = \mathbf{b} \tag{4}$$

We define matrix addition $A + B = C$ so that Equations (3) and (4) are the same. This is clearly the case if *corresponding elements are added*. Clearly, addition is only possible if A and B are the same size, in which case we say that A and B are *compatible for addition*.

Multiplication of Matrix by Scalar

Multiplying a matrix A by a scalar integer k is equivalent to adding A to itself k times; but this adds each element of A to itself k times, that is, each element of A is multiplied by k. Thus, $kA = [ka_{ij}]$. This rule followed naturally from the rule for adding matrices. We use it as the *definition* of multiplication, when k is *any* real scalar.

Matrix Multiplication

To define matrix multiplication as a natural extension of matrix-vector multiplication, consider first the 2×2 case. Let

$$A = \begin{bmatrix} a_{11} & a_{12} \\ a_{21} & a_{22} \end{bmatrix}, \quad B = \begin{bmatrix} b_{11} & b_{12} \\ b_{21} & b_{22} \end{bmatrix}, \quad C = \begin{bmatrix} c_{11} & c_{12} \\ c_{21} & c_{22} \end{bmatrix} \quad \text{and} \quad s = \begin{bmatrix} s_1 \\ s_2 \end{bmatrix}$$

If $r = Bs$, then $Ar = A(Bs)$; we should like to be able to write the last expression as $(AB)s = Cs$, say, where $C = AB$ is the product of A and B. To do this, we must define the product C so that it satisfies $A(Bs) = Cs$ for any value of s. But

$$Bs = \begin{bmatrix} b_{11}s_1 + b_{12}s_2 \\ b_{21}s_1 + b_{22}s_2 \end{bmatrix}$$

so that

$$A(Bs) = \begin{bmatrix} a_{11}(b_{11}s_1 + b_{12}s_2) + a_{12}(b_{21}s_1 + b_{22}s_2) \\ a_{21}(b_{11}s_1 + b_{12}s_2) + a_{22}(b_{21}s_1 + b_{22}s_2) \end{bmatrix}$$

$$= \begin{bmatrix} (a_{11}b_{11} + a_{12}b_{21})s_1 + (a_{11}b_{12} + a_{12}b_{22})s_2 \\ (a_{21}b_{11} + a_{22}b_{21})s_1 + (a_{21}b_{12} + a_{22}b_{22})s_2 \end{bmatrix}$$

This must be the same for any s, as

$$Cs = \begin{bmatrix} c_{11}s_1 + c_{12}s_2 \\ c_{21}s_1 + c_{22}s_2 \end{bmatrix}$$

and this will be the case if we choose

$$c_{11} = a_{11}b_{11} + a_{12}b_{21}; \quad c_{12} = a_{11}b_{12} + a_{12}b_{22};$$
$$c_{21} = a_{21}b_{11} + a_{22}b_{21}; \quad c_{22} = a_{21}b_{12} + a_{22}b_{22}$$

We generalise this to the case where A is $m \times n$, B is $n \times l$, C is $m \times l$ and s is an l-component vector. We shall use the notation $(\mathbf{p})_k$ for the kth component of the vector $\mathbf{p}$. Then the kth component of the vector Bs is

$$(Bs)_k = \sum_{j=1}^{l} b_{kj}s_j$$

and so the ith component of the vector $A(Bs)$ is

$$(A(Bs))_i = \sum_{k=1}^{n} a_{ik}(Bs)_k = \sum_{k=1}^{n} a_{ik} \left(\sum_{j=1}^{l} b_{kj}s_j \right)$$

$$= \sum_{j=1}^{l} \left(\sum_{k=1}^{n} a_{ik}b_{kj} \right) s_j,$$

where we have reversed the order of summation. This must be the same for every s, as

$$(Cs)_i = \sum_{j=1}^{l} c_{ij}s_j$$

which will be the case if

$$c_{ij} = \sum_{k=1}^{n} a_{ik}b_{kj}, \qquad j = 1, 2, \ldots, l$$

We want this to be true for all components of Cs, so the last equation must also be true for $i = 1, 2, \ldots, m$.

In words, the rule for multiplication is 'the element in the ith row and jth column of the product AB is formed by multiplying elements in the ith row of A by corresponding elements in the jth column of B and adding.

The matrix A is *compatible* for multiplication by B if the number of columns of A is the same as the number of rows of B. The sizes of A, B and C are illustrated by

$$A \quad \times \quad B \quad = \quad C$$
$$m \times n, \quad n \times l, \quad m \times l$$

When $l = 1$, B is an n-component vector and the product AB is an m-component vector.

It should be checked (from the rules) that, if the matrices are compatible for the operations involved,

(1) $B + C = C + B$ (commutativity of addition)
(2) $(A + B) + C = A + (B + C)$ (associativity of addition)
(3) $(AB)C = A(BC)$ (associativity of multiplication)
(4) If O is the null matrix (every element zero), then

$$A + O = O + A = A \quad \text{and} \quad OA = AO = O$$

Note especially that:

(a) The vanishing of the product AB does not imply that A or B is zero, since, for example,

$$\begin{bmatrix} 1 & 0 \\ 0 & 0 \end{bmatrix}\begin{bmatrix} 0 & 0 \\ 0 & 1 \end{bmatrix} = \begin{bmatrix} 0 & 0 \\ 0 & 0 \end{bmatrix}$$

(b) Similarly, $A\mathbf{r} = \mathbf{0}$ (the zero vector) does not imply that $A = O$ or $\mathbf{r} = \mathbf{0}$, since, for example,

$$\begin{bmatrix} 1 & -1 \\ -1 & 1 \end{bmatrix}\begin{bmatrix} 1 \\ 1 \end{bmatrix} = \begin{bmatrix} 0 \\ 0 \end{bmatrix}$$

This means that $A\mathbf{r} = A\mathbf{s}$ and $A \neq O$ does not imply that $\mathbf{r} = \mathbf{s}$, since although $A\mathbf{r} = A\mathbf{s}$ means that $A(\mathbf{r} - \mathbf{s}) = \mathbf{0}$, this does not imply that $\mathbf{r} - \mathbf{s}$ is the zero vector.

EXERCISES 4.4.1

1 Find all possible products of pairs of the following matrices:

$$A = [1 \quad 2 \quad 3]; \quad B = \begin{bmatrix} 1 & 2 & 3 \\ 2 & 3 & 4 \end{bmatrix}; \quad C = \begin{bmatrix} 1 \\ 2 \\ 3 \end{bmatrix}; \quad D = \begin{bmatrix} 1 & 2 \\ 3 & 4 \end{bmatrix}; \quad E = \begin{bmatrix} 1 \\ 2 \end{bmatrix}.$$

2 Let

$$A = \begin{bmatrix} \frac{1}{3} & \frac{1}{3} & \frac{1}{3} \\ \frac{1}{3} & \frac{1}{3} & \frac{1}{3} \\ \frac{1}{3} & \frac{1}{3} & \frac{1}{3} \end{bmatrix}$$

Find A^n for all positive integers n.

4.5 SQUARE MATRICES

These are, as we have already seen, matrices with the same number of rows and columns. If A and B are both $n \times n$, then both products AB and BA can be formed, but these are not generally the same. For example,

$$\begin{bmatrix} 1 & 2 \\ 3 & 4 \end{bmatrix} \begin{bmatrix} 2 & 3 \\ 4 & 5 \end{bmatrix} = \begin{bmatrix} 10 & 13 \\ 22 & 29 \end{bmatrix}$$

while

$$\begin{bmatrix} 2 & 3 \\ 4 & 5 \end{bmatrix} \begin{bmatrix} 1 & 2 \\ 3 & 4 \end{bmatrix} = \begin{bmatrix} 11 & 16 \\ 19 & 17 \end{bmatrix}$$

Thus, matrix multiplication is not commutative. It is useful to say that A *pre-multiplies* B, to form AB, or that A *post-multiplies* B, to form BA. We must take care, when multiplying through an equation by a matrix A, that each term is pre-multiplied by A or that each term is post-multiplied by A.

Identity Matrix

Is there a matrix I such that, for any matrix A, $AI = IA = A$? It is easy to check that the matrix I is the matrix (of size compatible for multiplication) consisting of 1 in each diagonal position and zeros in every other position. The matrix I is called the *identity matrix*. It should be noted also that for any vector $\mathbf{x}$, which is compatible for multiplication, $I\mathbf{x} = \mathbf{x}$.

Inverse Matrix

For an equation in real numbers, such as $3x = 5$, we write the solution as $x = \frac{5}{3}$; we might have written it as $\frac{1}{3} \times 5$, where $\frac{1}{3}$ is the inverse of 3. Multiplying both sides of the equation $3x = 5$ by this inverse, we obtain

$$\tfrac{1}{3}3x = \tfrac{1}{3}5$$

and it is clear that the product of 3 and its inverse $\tfrac{1}{3}$ is 1, so that the left-hand side becomes $1x$ or just x, and the right-hand side gives the required solution.

If we have a matrix equation $Ax = b$, we should like to find a matrix B such that $BA = I$, for then we should have

$$BAx = Bb$$

and the left-hand side becomes $Ix = x$, and so we have the solution $x = Bb$. If such a B exists, it is called the *left inverse* of A. If B is also unique, then we can deduce that $AB = I$ also. For

$$(I - AB + B)A = A - (AB)A + BA$$
$$= A - A(BA) + BA$$
$$= A - AI + I$$
$$= A - A + I$$
$$= I$$

This shows that $I - AB + B$ is the left inverse of A, but since this is uniquely B, we must have

$$I - AB + B = B$$

giving $I - AB = 0$, so that $AB = I$. In this case, B is also the right inverse of A; we write $B = A^{-1}$ and call this simply the *inverse* of A. Thus, $AA^{-1} = A^{-1}A = I$.

Note that because $A^{-1}B \neq BA^{-1}$, in general, we cannot use the notation

$$\frac{B}{A}$$

since this does not specify the order of multiplication.

So far we do not even know whether a matrix has an inverse. We now investigate this problem, starting by trying to find the inverse of a 2×2 example.

Examples 4.5.1

1. Find the inverse of the matrix

$$\begin{bmatrix} 1 & 0 \\ 1 & 1 \end{bmatrix}$$

Suppose the inverse, if it exists, is

$$\begin{bmatrix} e & g \\ f & h \end{bmatrix}$$

so that we must have

$$\begin{bmatrix} 1 & 0 \\ 1 & 1 \end{bmatrix}\begin{bmatrix} e & g \\ f & h \end{bmatrix} = \begin{bmatrix} 1 & 0 \\ 0 & 1 \end{bmatrix} \qquad (1)$$

It is useful to split this matrix equation into two separate ones, each containing just two of the unknowns:

$$\begin{bmatrix} 1 & 0 \\ 1 & 1 \end{bmatrix}\begin{bmatrix} e \\ f \end{bmatrix} = \begin{bmatrix} 1 \\ 0 \end{bmatrix} \quad \text{and} \quad \begin{bmatrix} 1 & 0 \\ 1 & 1 \end{bmatrix}\begin{bmatrix} g \\ h \end{bmatrix} = \begin{bmatrix} 0 \\ 1 \end{bmatrix} \qquad (2)$$

Equating corresponding elements on the left and right of the equations (2), we obtain

$$e = 1, e + f = 0, g = 0 \quad \text{and} \quad g + h = 1$$

which give $f = -1$ and $h = 1$. (Check that Equation (1) gives the same result.) The inverse is therefore

$$\begin{bmatrix} 1 & 0 \\ -1 & 1 \end{bmatrix}$$

This can be checked by post-multiplying by the original matrix.

2. Find the inverse of the matrix

$$\begin{bmatrix} 2 & 6 \\ 1 & 3 \end{bmatrix}$$

Letting the inverse be as in Example 1, we obtain the equations

$$\begin{bmatrix} 2 & 6 \\ 1 & 3 \end{bmatrix}\begin{bmatrix} e \\ f \end{bmatrix} = \begin{bmatrix} 1 \\ 0 \end{bmatrix} \quad \text{and} \quad \begin{bmatrix} 2 & 6 \\ 1 & 3 \end{bmatrix}\begin{bmatrix} g \\ h \end{bmatrix} = \begin{bmatrix} 0 \\ 1 \end{bmatrix}$$

which yield the equations

$$2e + 6f = 1, e + 3f = 0, 2g + 6h = 0 \quad \text{and} \quad g + 3h = 1$$

These two pairs of equations are inconsistent and therefore have no solution. We deduce that there is no inverse in this case, and we call the matrix *singular*.

It is clear that not all matrices possess an inverse; we investigate the situation further by trying to find the inverse of the matrix

$$A = \begin{bmatrix} a & b \\ c & d \end{bmatrix}$$

Let it be

$$\begin{bmatrix} e & g \\ f & h \end{bmatrix}$$

if it exists. Then

$$\begin{bmatrix} a & b \\ c & d \end{bmatrix}\begin{bmatrix} e \\ f \end{bmatrix} = \begin{bmatrix} 1 \\ 0 \end{bmatrix} \quad \text{and} \quad \begin{bmatrix} a & b \\ c & d \end{bmatrix}\begin{bmatrix} g \\ h \end{bmatrix} = \begin{bmatrix} 0 \\ 1 \end{bmatrix}$$

The condition for each of these equations to have a unique solution is the non-vanishing of the determinant of coefficients,

$$\det A = \begin{vmatrix} a & b \\ c & d \end{vmatrix}$$

which is then the condition for the non-singularity of the matrix (that is, for the matrix to have an inverse).

For an $n \times n$ matrix A, the condition $\det A \neq 0$ also guarantees that A is non-singular and that the equation $A\mathbf{x} = \mathbf{b}$ has a unique solution.

We now show how to compute the inverses, when they exist, of 3×3 or larger matrices. First, we repeat the solution of Example 4.3.1, using a more compact notation, called the *augmented matrix*, which consists of the matrix of coefficients augmented by the right-hand side vector,

$$\begin{bmatrix} 1 & 5 & 3 & a \\ 1 & 2 & k & b \\ 5 & 1 & -k & c \end{bmatrix}$$

(1)
(2)
(3)

Each row of this matrix represents an equation; each operation we performed before, which consisted of subtracting a multiple of one equation from another, is now replaced by a corresponding operation on the rows of the augmented matrix. As in the previous solution, we show the operations by the numbers on the right. The solution now proceeds as follows:

$$\begin{bmatrix} 1 & 5 & 3 & a \\ 0 & -3 & k-3 & b-a \\ 0 & -24 & -(k+15) & c-5a \end{bmatrix}$$

$$(1)$$
$$(2)' = (2) - (1)$$
$$(3)' = (3) - 5(1)$$

$$\begin{bmatrix} 1 & 5 & 3 & a \\ 0 & -3 & k-3 & b-a \\ 0 & 0 & 9(1-k) & 3a-8b+c \end{bmatrix}$$

$$(1)$$
$$(2)'$$
$$(3)'' = (3)' - 8(2)'$$

The only difference in this solution is the notation; we recall that the augmented matrix represents three equations, which we now solve by back-substitution, as before.

The operations we used above are called *elementary row operations*, and we use them now to compute the inverse of the 3×3 matrix

$$A = \begin{bmatrix} 1 & 2 & 3 \\ 2 & 3 & 4 \\ 3 & 4 & 5 \end{bmatrix}$$

Let the inverse be

$$\begin{bmatrix} a_1 & b_1 & c_1 \\ a_2 & b_2 & c_2 \\ a_3 & b_3 & c_3 \end{bmatrix}$$

This must satisfy the equation

$$\begin{bmatrix} 1 & 2 & 3 \\ 2 & 3 & 4 \\ 3 & 4 & 5 \end{bmatrix} \begin{bmatrix} a_1 & b_1 & c_1 \\ a_2 & b_2 & c_2 \\ a_3 & b_3 & c_3 \end{bmatrix} = \begin{bmatrix} 1 & 0 & 0 \\ 0 & 1 & 0 \\ 0 & 0 & 1 \end{bmatrix}$$

which we can write in separated form as

$$A \begin{bmatrix} a_1 \\ a_2 \\ a_3 \end{bmatrix} = \begin{bmatrix} 1 \\ 0 \\ 0 \end{bmatrix}, \quad A \begin{bmatrix} b_1 \\ b_2 \\ b_3 \end{bmatrix} = \begin{bmatrix} 0 \\ 1 \\ 0 \end{bmatrix} \quad \text{and} \quad A \begin{bmatrix} c_1 \\ c_2 \\ c_3 \end{bmatrix} = \begin{bmatrix} 0 \\ 0 \\ 1 \end{bmatrix}$$

as we can see by multiplying out. We need to solve the equation $Ax = b$ for the three right-hand sides

$$\begin{bmatrix} 1 \\ 0 \\ 0 \end{bmatrix}, \begin{bmatrix} 0 \\ 1 \\ 0 \end{bmatrix} \quad \text{and} \quad \begin{bmatrix} 0 \\ 0 \\ 1 \end{bmatrix}$$

We therefore write our augmented matrix to include all these right-hand sides as

$$\begin{bmatrix} 1 & 2 & 3 & 1 & 0 & 0 \\ 2 & 3 & 4 & 0 & 1 & 0 \\ 3 & 4 & 5 & 0 & 0 & 1 \end{bmatrix}$$

Now, when we perform row operations, we are performing operations on three sets of equations at the same time. This is obviously more economical than treating each set separately; for the operations must be identical for each set, since they depend only on the left-hand sides of the equations. We proceed as follows:

$$\begin{bmatrix} 1 & 2 & 3 & 1 & 0 & 0 \\ 0 & -1 & -2 & -2 & 1 & 0 \\ 0 & -2 & -4 & -3 & 0 & 1 \end{bmatrix}$$

$$\rightarrow \begin{bmatrix} 1 & 2 & 3 & 1 & 0 & 0 \\ 0 & -1 & -2 & -2 & 1 & 0 \\ 0 & 0 & 0 & 1 & -2 & 1 \end{bmatrix}$$

At this stage it is clear that the equations have no solution, since the left-hand side of the last equation has vanished. The determinant of coefficients is also zero, so that the matrix A is singular and has no inverse. In order to illustrate a complete inversion calculation, we alter the last element of A to 6, which makes A non-singular. The calculation now is

$$\begin{bmatrix} 1 & 2 & 3 & 1 & 0 & 0 \\ 2 & 3 & 4 & 0 & 1 & 0 \\ 3 & 4 & 6 & 0 & 0 & 1 \end{bmatrix}$$

$$\rightarrow \begin{bmatrix} 1 & 2 & 3 & 1 & 0 & 0 \\ 0 & -1 & -2 & -2 & 1 & 0 \\ 0 & -2 & -3 & -3 & 0 & 1 \end{bmatrix}$$

$$\rightarrow \begin{bmatrix} 1 & 2 & 3 & 1 & 0 & 0 \\ 0 & -1 & -2 & -2 & 1 & 0 \\ 0 & 0 & 1 & 1 & -2 & 1 \end{bmatrix}$$

whence we could solve for each of the three right-hand sides. However, we choose to carry the reduction further:

$$\begin{bmatrix} 1 & 2 & 0 & -2 & 6 & -3 \\ 0 & -1 & 0 & 0 & -3 & 2 \\ 0 & 0 & 1 & 1 & -2 & 1 \end{bmatrix}$$

$$\rightarrow \begin{bmatrix} 1 & 0 & 0 & -2 & 0 & 1 \\ 0 & 1 & 0 & 0 & 3 & -2 \\ 0 & 0 & 1 & 1 & -2 & 1 \end{bmatrix}$$

This augmented matrix represents the equation

$$\begin{bmatrix} 1 & 0 & 0 \\ 0 & 1 & 0 \\ 0 & 0 & 1 \end{bmatrix} \begin{bmatrix} a_1 & b_1 & c_1 \\ a_2 & b_2 & c_2 \\ a_3 & b_3 & c_3 \end{bmatrix} = \begin{bmatrix} -2 & 0 & 1 \\ 0 & 3 & -2 \\ 1 & -2 & 1 \end{bmatrix}$$

but since the first matrix is just the identity matrix, we obtain

$$\begin{bmatrix} a_1 & b_1 & c_1 \\ a_2 & b_2 & c_2 \\ a_3 & b_3 & c_3 \end{bmatrix} = \begin{bmatrix} -2 & 0 & 1 \\ 0 & 3 & -3 \\ 1 & -2 & 1 \end{bmatrix}$$

which is the required inverse of A.

A summary of the method, which works for a matrix of any size, is: To invert a matrix, augment it with the identity matrix and use elementary row operations to reduce the original matrix to the identity matrix. If this cannot be done (for example, if a zero row occurs), the matrix is singular; otherwise, the inverse is left in the place originally occupied by the identity matrix.

This method is just a systematic way of solving $AB = I$ to find B.

Example 4.5.2

Find the inverse of the matrix

$$\begin{bmatrix} 1 & 3 & 1 \\ 2 & 6 & 3 \\ -1 & 2 & 5 \end{bmatrix}$$

Following the method above, we obtain

$$\begin{bmatrix} 1 & 3 & 1 & 1 & 0 & 0 \\ 2 & 6 & 3 & 0 & 1 & 0 \\ -1 & 2 & 5 & 0 & 0 & 1 \end{bmatrix}$$

$$\rightarrow \begin{bmatrix} 1 & 3 & 1 & 1 & 0 & 0 \\ 0 & 0 & 1 & -2 & 1 & 0 \\ 0 & 5 & 6 & 1 & 0 & 1 \end{bmatrix}$$

It appears that we can proceed no further because of the zero element in the second row and column. However, we interchange the whole of the second and third rows and proceed as usual:

$$\begin{bmatrix} 1 & 3 & 1 & 1 & 0 & 0 \\ 0 & 5 & 6 & 1 & 0 & 1 \\ 0 & 0 & 1 & -2 & 1 & 0 \end{bmatrix}$$

$$\rightarrow \begin{bmatrix} 1 & 3 & 0 & 3 & -1 & 0 \\ 0 & 5 & 0 & 13 & -6 & 1 \\ 0 & 0 & 1 & -2 & 1 & 0 \end{bmatrix}$$

$$\rightarrow \begin{bmatrix} 1 & 0 & 0 & -\frac{24}{5} & \frac{13}{5} & -\frac{3}{5} \\ 0 & 1 & 0 & \frac{13}{5} & -\frac{6}{5} & \frac{1}{5} \\ 0 & 0 & 1 & -2 & 1 & 0 \end{bmatrix}$$

and the inverse appears in the last three columns.

Transpose and Inverse of Matrix Products

Let $A = [a_{ij}]$ be an $m \times n$ matrix. Then the *transpose* of A is the $n \times m$ matrix $A^T = [a_{ji}]$. Thus, the transpose of A is obtained by interchanging for every i and j the element in the ith row and jth column and the element in the jth row and ith column. For example,

$$\begin{bmatrix} 1 & 2 & 3 \\ 4 & 5 & 6 \end{bmatrix}^T = \begin{bmatrix} 1 & 4 \\ 2 & 5 \\ 3 & 6 \end{bmatrix}$$

We now deduce a few useful properties of the transpose and inverse of a product of matrices.

(a) $(AB)^T = B^T A^T$: We assume that $B = [b_{ij}]$ has n rows and l columns, so that the product AB exists. Then the ijth element of AB is

$$(AB)_{ij} = \sum_{k=1}^{n} a_{ik} b_{kj}$$

so the ijth element of $(AB)^T$, which we obtain by swapping i and j, is

$$((AB)^T)_{ij} = \sum_{k=1}^{n} a_{jk} b_{ki}$$

$$= \sum_{k=1}^{n} b_{ki} a_{jk}$$

$$= (B^T A^T)_{ij}$$

(b) $(AB)^{-1} = B^{-1} A^{-1}$, where A and B are non-singular square matrices: Let $C = (AB)^{-1}$; then C must satisfy

$$ABC = I$$

Pre-multiplying both sides of this equation by A^{-1}, we obtain

$$A^{-1} ABC = A^{-1} I$$

and, replacing $A^{-1}A$ by I, then IB by B, and $A^{-1}I$ by A^{-1}, we find

$$BC = A^{-1}$$

119

Now multiplying both sides by B^{-1} and simplifying, we obtain

$$C = B^{-1}A^{-1}$$

(c) $(A^T)^{-1} = (A^{-1})^T$, where A is a non-singular square matrix:

$$(AA^{-1})^T = I^T = I$$

so, using rule (a) above, we have

$$(A^{-1})^T A^T = I$$

but this shows that $(A^{-1})^T$ is the inverse of A^T, as required.

EXERCISES 4.5.1

1 Check which of the following matrices are non-singular, and find their inverses where possible:

$$A = \begin{bmatrix} 1 & 2 \\ 3 & 4 \end{bmatrix}; \quad B = \begin{bmatrix} 5 & 0 & 3 \\ 2 & -1 & 0 \\ 0 & 1 & -2 \end{bmatrix}; \quad C = \begin{bmatrix} 3 & -1 & -2 \\ 0 & 2 & -1 \\ 3 & -5 & 0 \end{bmatrix};$$

$$D = \begin{bmatrix} 0 & 1 & 2 \\ -1 & 0 & 1 \\ 2 & 3 & 1 \end{bmatrix}.$$

2 Find the inverse of the matrix

$$\begin{bmatrix} a & b \\ c & d \end{bmatrix}$$

where $ad \neq bc$.

3 Show that if A, B, C are non-singular matrices, then

$$(ABC)^{-1} = C^{-1}B^{-1}A^{-1}.$$

4 Show that the *upper triangular* matrix

$$\begin{bmatrix} 1 & a & b \\ 0 & 1 & c \\ 0 & 0 & 1 \end{bmatrix}$$

has an inverse of the same form. Find values of d, e, f, g, h, i to satisfy

$$\begin{bmatrix} 1 & a & b \\ 0 & 1 & c \\ 0 & 0 & 1 \end{bmatrix} \begin{bmatrix} d & 0 & 0 \\ e & g & 0 \\ f & h & i \end{bmatrix} = \begin{bmatrix} 14 & 8 & 3 \\ 8 & 5 & 2 \\ 3 & 2 & 1 \end{bmatrix}$$

Find the inverse of these three matrices.

4.6 NUMERICAL SOLUTION OF LINEAR EQUATIONS

Gaussian Elimination

Many methods have been devised for the solution of linear equations, and a large proportion of these are variants of the classical method of Gaussian Elimination. We have, in fact, already followed the most straightforward form of this method whenever we have solved linear equations. We now describe this method, using the augmented matrix to abbreviate the equations.

For the equation $A\mathbf{x} = \mathbf{b}$, where $A = [a_{ij}]$, $\mathbf{x} = [x_1, x_2, \ldots, x_n]^T$ and $\mathbf{b} = [b_1, b_2, \ldots, b_n]^T$, we eliminate the variables in the natural order, that is, in the order $x_1, x_2, \ldots$, by subtracting suitable multiples of one equation (called the *pivotal equation*) of the augmented matrix from the other equations. We start with the first equation as pivotal equation to eliminate x_1 from the other $n-1$ equations, then use the second to eliminate x_2 from the last $n-2$ equations, and so on. At the end of this elimination process, we are left with an equation containing only the variable x_n, from which we find x_n. We then substitute this back into the previous equation, which enables us to find x_{n-1}, and so on back through all the equations; this process is called *back-substitution*. We illustrate the method with an example.

Example 4.6.1

Solve the equations

$$x + 2y + 3z = 1$$

$$2x + 3y + 5z = 0$$

$$3x + 4y + 5z = 0$$

We write in augmented matrix form and eliminate the variables as described above:

$$\begin{bmatrix} 1 & 2 & 3 & 1 \\ 2 & 3 & 5 & 0 \\ 3 & 4 & 5 & 0 \end{bmatrix}$$

$$\rightarrow \begin{bmatrix} 1 & 2 & 3 & 1 \\ 0 & -1 & -1 & -2 \\ 0 & -2 & -4 & -3 \end{bmatrix}$$

$$\rightarrow \begin{bmatrix} 1 & 2 & 3 & 1 \\ 0 & -1 & -1 & -2 \\ 0 & 0 & -2 & 1 \end{bmatrix}$$

The last row represents the equation $-2z = 1$, so we find $z = -\frac{1}{2}$. Substituting this into the second equation, $-y - z = -2$, we find $y = \frac{5}{2}$, and putting these values for y and z into the first equation, $x + 2y + 3z = 1$, we find $x = -\frac{1}{2}$.

Now change the example slightly by replacing the coefficient of y in the second equation by 4, so that the augmented matrix is

$$\begin{bmatrix} 1 & 2 & 3 & 1 \\ 2 & 4 & 5 & 0 \\ 3 & 4 & 5 & 0 \end{bmatrix}$$

Proceeding with the elimination, we find

$$\begin{bmatrix} 1 & 2 & 3 & 1 \\ 0 & 0 & -1 & -2 \\ 0 & -2 & -4 & -3 \end{bmatrix}$$

We can no longer proceed with the elimination, because the y coefficient in the second equation has become zero (cf. Example 4.5.2). To proceed, we just interchange the second and third rows:

$$\begin{bmatrix} 1 & 2 & 3 & 1 \\ 0 & -2 & -4 & -3 \\ 0 & 0 & -1 & -2 \end{bmatrix}$$

and complete the solution with the back-substitution.

It is clear that the straightforward version of Gaussian Elimination must be modified to avoid this situation. With more equations, a choice will have to be made as to which equations to interchange. The usual choice, and one which also restricts the growth of rounding errors (the errors made in arithmetic operations because only a finite number of decimal places can be kept), is to choose the *pivot* (that is, the coefficient of the variable currently being eliminated) of maximum absolute value. When we came to eliminate y in the last example, the pivot turned out to be zero, so we chose instead -2 as the pivot from the last row; this corresponds to the choice of pivot of maximum size.

Example 4.6.2

We repeat the solution of Example 4.6.1, using maximum pivoting:

$$\begin{bmatrix} 1 & 2 & 3 & 1 \\ 2 & 3 & 5 & 0 \\ 3 & 4 & 5 & 0 \end{bmatrix}$$

The possible pivots for eliminating x are 1, 2 and 3, of which the last is biggest, so we interchange the first and third rows:

$$\begin{bmatrix} 3 & 4 & 5 & 0 \\ 2 & 3 & 5 & 0 \\ 1 & 2 & 3 & 1 \end{bmatrix}$$

and proceed to eliminate x:

$$\begin{bmatrix} 3 & 4 & 5 & 0 \\ 0 & \frac{1}{3} & \frac{5}{3} & 0 \\ 0 & \frac{2}{3} & \frac{4}{3} & 1 \end{bmatrix}$$

The choice of pivots to eliminate y is $\frac{1}{3}$ and $\frac{2}{3}$, so we choose $\frac{2}{3}$ and proceed as usual:

$$\begin{bmatrix} 3 & 4 & 5 & 0 \\ 0 & \frac{2}{3} & \frac{4}{3} & 1 \\ 0 & \frac{1}{3} & \frac{5}{3} & 0 \end{bmatrix}$$

$$\rightarrow \begin{bmatrix} 3 & 4 & 5 & 0 \\ 0 & \frac{2}{3} & \frac{4}{3} & 1 \\ 0 & 0 & 1 & -\frac{1}{2} \end{bmatrix},$$

and complete the solution with the back-substitution.

There are several points to note about this method, which is called Gaussian Elimination with pivoting.

(1) It might be thought that the arithmetic is unnecessarily complicated, since fractions could be avoided by eliminating in a different way. However, remember that this is only an illustrative example; in a real problem the numbers could be long decimals and there could be hundreds of equations. In such situations, where a computer would be needed for the solution, Gaussian Elimination with pivoting is established as a reliable tool.

(2) If at some stage all the possible pivots vanish, the matrix of coefficients is singular and so a unique solution does not exist. There may still be an infinity of possible solutions, but this is not usually of interest in a practical case.

(3) One might be tempted to take the elimination of variables further, as

when inverting a matrix. For Example 4.6.2 this would give

$$\begin{bmatrix} 3 & 4 & 5 & 0 \\ 0 & 1 & 2 & \frac{3}{2} \\ 0 & 0 & 1 & -\frac{1}{2} \end{bmatrix}$$

$$\rightarrow \begin{bmatrix} 3 & 4 & 0 & \frac{5}{2} \\ 0 & 1 & 0 & \frac{5}{2} \\ 0 & 0 & 1 & -\frac{1}{2} \end{bmatrix}$$

$$\rightarrow \begin{bmatrix} 3 & 0 & 0 & -\frac{15}{2} \\ 0 & 1 & 0 & \frac{5}{2} \\ 0 & 0 & 1 & -\frac{1}{2} \end{bmatrix}$$

$$\rightarrow \begin{bmatrix} 1 & 0 & 0 & -\frac{5}{2} \\ 0 & 1 & 0 & \frac{5}{2} \\ 0 & 0 & 1 & -\frac{1}{2} \end{bmatrix}$$

which yields the solution immediately. However, it turns out that the number of arithmetic operations $(+, -, \times, /)$ needed is more than that for the back-substitution method.

EXERCISE 4.6.1

Use Gaussian Elimination with and without pivoting to solve the equations

$$0.00137x + 0.859y = 1.00$$

$$0.962x + 0.0149y = 1.00$$

Use a calculator, and round *every* arithmetic operation to three figures.

MISCELLANEOUS EXERCISES 4

1 Find the quadratic which interpolates $\sin x$ at $x = 0, \frac{\pi}{2}, \pi$. (*Hint*: Let the quadratic be $p(x) = ax^2 + bx + c$; then the interpolating conditions $p(0) = \sin 0$, $p(\frac{\pi}{2}) = \sin \frac{\pi}{2}$ and $p(\pi) = \sin \pi$ give three simultaneous equations for a, b, c.) Use the result to estimate $\sin \frac{\pi}{3}$.
2 Figure 4.1 illustrates the flow of traffic in vehicles per hour at a roundabout. Assuming that the flow into a junction equals the flow out of it, write down four simultaneous equations for x_1, x_2, x_3, x_4 and solve them for the case when $x_4 = 100$ vehicles per hour.
3 Find the currents i_1, i_2, i_3 in the electrical network shown in Figure 4.2,

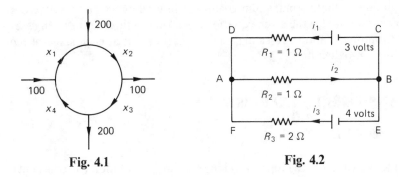

Fig. 4.1   Fig. 4.2

which satisfy the equations

$$i_1 + i_3 = i_2 \text{ (junction A or B)}$$

$$i_1 + i_2 = 3 \text{ (circuit ABCD)}$$

$$i_2 + 2i_3 = 4 \text{ (circuit ABEF)}$$

4 For a set of points $(x_1, y_1), (x_2, y_2), \ldots, (x_n, y_n)$ the least squares regression line is given by the linear function $f(x) = a_0 + a_1 x$ that minimises the sum of the squares of the errors, $\sum_{i=1}^{n} \{y_i - f(x_i)\}^2$. The values of a, b are obtained from the solution of

$$(X^T X) \begin{bmatrix} a_0 \\ a_1 \end{bmatrix} = X^T y$$

where

$$X = \begin{bmatrix} 1 & x_1 \\ 1 & x_2 \\ \vdots & \vdots \\ 1 & x_n \end{bmatrix}, \qquad y = \begin{bmatrix} y_1 \\ y_2 \\ \vdots \\ y_n \end{bmatrix}$$

Find the least squares regression line for the points $(1, 1)$, $(2, 2)$, $(3, 4)$, $(4, 4)$, $(5, 6)$, and sketch it on a graph, showing also the data points.

5 Suppose the first equation in Exercise 4.6.1 had been multiplied by 1000; the equations would then be

$$1.37x + 859y = 1000$$

$$0.962x + 0.0149y = 1.00$$

Solving these equations by Gaussian Elimination with maximum pivoting would require 1.37 to be the pivot now, rather than 0.962, and the solution would proceed as in Exercise 4.6.1 *without* pivoting. Thus, the choice of pivots depends on the relative scaling of the equations. One way of removing this rather artificial dependence is to scale each equation

equation so that its maximum coefficient in modulus is unity. Carry out this procedure for the given equations, working to three figures. (In cases with more equations, rescaling must be carried out at each stage of the Gaussian reduction.)

4.7 ANSWERS TO EXERCISES

Exercises 4.1.1

1 (i) $(2, 1)$, (ii) $(-2\lambda, \lambda)$ for any value of λ, (iii) no solution (inconsistent), (iv) $(1 - 2\lambda, \lambda)$ for any value of λ.
2 (i) -2, (ii) 11, (iii) -7, (iv) 1.

Exercises 4.2.1

1 (i) -240, (ii) 0.

$$\mathbf{2} \quad \begin{vmatrix} -x & 1 & 0 \\ 1 & -x & 1 \\ 0 & 1 & -x \end{vmatrix} = 0 \Rightarrow -x \begin{vmatrix} -x & 1 \\ 1 & -x \end{vmatrix} - 1 \begin{vmatrix} 1 & 1 \\ 0 & -x \end{vmatrix} = 0$$

$$\Rightarrow -x^2 + 2x = 0 \Rightarrow x = 0, \ \pm\sqrt{2}.$$

Exercises 4.3.1

1 (i) $(1, -1, 0)$,

(ii) $\left(\dfrac{17 + \lambda}{11}, \dfrac{5\lambda - 3}{11}, \lambda \right)$ for any value of λ,

(iii) no solution (inconsistent), (iv) $(1 - \lambda - \mu, \lambda, \mu)$ for any values of λ, μ.
2 Labelling the given equations (1), (2) and (3), we find

$$x + y + z = 3 \qquad\qquad (1)$$
$$y + z = 2 \qquad\qquad (4) = (2) - (1)$$
$$x = 1 \qquad\qquad (5) = (1) - (4)$$
$$(a - 1)y + (b - 1)z = 0 \qquad (6) = (3) - (1)$$
$$(b - a)z = -2(a - 1) \quad (7) = (6) - (a - 1) \times (4)$$

Then solutions are: (i) $(1, p, 2 - p)$ for any value of p, (ii) $(1, 2, 0)$, (iii) $(1, 0, 2)$, (iv) no solution,

(v) $\left(1, \dfrac{2(b - 1)}{b - a}, \dfrac{-2(a - 1)}{b - a} \right)$.

126

Exercises 4.4.1

1 $AC = [14]$, $BC = \begin{bmatrix} 14 \\ 20 \end{bmatrix}$, $CA = \begin{bmatrix} 1 & 2 & 3 \\ 2 & 4 & 6 \\ 3 & 6 & 9 \end{bmatrix}$, $DB = \begin{bmatrix} 5 & 8 & 11 \\ 11 & 18 & 25 \end{bmatrix}$,

$DD = D^2 = \begin{bmatrix} 7 & 10 \\ 15 & 22 \end{bmatrix}$, $DE = \begin{bmatrix} 5 \\ 11 \end{bmatrix}$, $EA = \begin{bmatrix} 1 & 2 & 3 \\ 2 & 4 & 6 \end{bmatrix}$.

2 $A^n = A$.

Exercises 4.5.1

1 $\det A = -2$, $A^{-1} = \begin{bmatrix} -2 & 1 \\ \frac{3}{2} & -\frac{1}{2} \end{bmatrix}$,

$\det B = 16$, $B^{-1} = \begin{bmatrix} \frac{1}{8} & \frac{3}{16} & \frac{3}{16} \\ \frac{1}{4} & -\frac{5}{8} & \frac{3}{8} \\ \frac{1}{8} & -\frac{5}{16} & -\frac{5}{16} \end{bmatrix}$,

$\det C = 0$, so C is singular,

$\det D = -3$, $D^{-1} = \begin{bmatrix} 1 & -\frac{5}{3} & -\frac{1}{3} \\ -1 & \frac{4}{3} & \frac{2}{3} \\ 1 & -\frac{2}{3} & -\frac{1}{3} \end{bmatrix}$.

2 Reduce the augmented matrix thus:

$$\begin{bmatrix} a & b & 1 & 0 \\ c & d & 0 & 1 \end{bmatrix} \rightarrow \begin{bmatrix} 1 & \dfrac{b}{a} & \dfrac{1}{a} & 0 \\ 0 & \dfrac{ad - bc}{a} & -\dfrac{c}{a} & 1 \end{bmatrix}$$

$$\rightarrow \begin{bmatrix} 1 & 0 & \dfrac{1}{a} - \dfrac{bc/a}{ad - bc} & -\dfrac{b}{ad - bc} \\ 0 & 1 & -\dfrac{c}{ad - bc} & \dfrac{a}{ad - bc} \end{bmatrix}$$

$$\rightarrow \begin{bmatrix} 1 & 0 & \dfrac{d}{ad - bc} & -\dfrac{b}{ad - bc} \\ 0 & 1 & -\dfrac{c}{ad - bc} & \dfrac{a}{ad - bc} \end{bmatrix}.$$

Thus the inverse is

$$\begin{bmatrix} d & -b \\ -c & a \end{bmatrix} \Big/ \begin{vmatrix} a & b \\ c & d \end{vmatrix}.$$

3 Using the result for the inverse of a product of two matrices,
$(A(BC))^{-1} = (BC)^{-1}A^{-1} = C^{-1}B^{-1}A^{-1}$.

4

$$\begin{bmatrix} 1 & a & b & 1 & 0 & 0 \\ 0 & 1 & c & 0 & 1 & 0 \\ 0 & 0 & 1 & 0 & 0 & 1 \end{bmatrix} \rightarrow \begin{bmatrix} 1 & a & 0 & 1 & 0 & -b \\ 0 & 1 & 0 & 0 & 1 & -c \\ 0 & 0 & 1 & 0 & 0 & 1 \end{bmatrix}$$

$$\rightarrow \begin{bmatrix} 1 & 0 & 0 & 1 & -a & ac-b \\ 0 & 1 & 0 & 0 & 1 & -c \\ 0 & 0 & 1 & 0 & 0 & 1 \end{bmatrix},$$

which shows that the inverse has the same form.

$$\begin{bmatrix} 1 & 2 & 3 \\ 0 & 1 & 2 \\ 0 & 0 & 1 \end{bmatrix} \begin{bmatrix} 1 & 0 & 0 \\ 2 & 1 & 0 \\ 3 & 2 & 1 \end{bmatrix} = \begin{bmatrix} 14 & 8 & 3 \\ 8 & 5 & 3 \\ 3 & 2 & 1 \end{bmatrix}.$$

Using the first part and the inverse of the transpose rule:

$$\begin{bmatrix} 1 & 2 & 3 \\ 0 & 1 & 2 \\ 0 & 0 & 1 \end{bmatrix}^{-1} = \begin{bmatrix} 1 & -2 & 1 \\ 0 & 1 & -2 \\ 0 & 0 & 1 \end{bmatrix}, \quad \begin{bmatrix} 1 & 0 & 0 \\ 2 & 1 & 0 \\ 3 & 2 & 1 \end{bmatrix}^{-1} = \begin{bmatrix} 1 & 0 & 0 \\ -2 & 1 & 0 \\ 1 & -2 & 0 \end{bmatrix}.$$

Using the inverse of the product rule:

$$\begin{bmatrix} 14 & 8 & 3 \\ 8 & 5 & 2 \\ 3 & 2 & 1 \end{bmatrix}^{-1} = \begin{bmatrix} 1 & 0 & 0 \\ -2 & 1 & 0 \\ 1 & -2 & 1 \end{bmatrix} \begin{bmatrix} 1 & -2 & 1 \\ 0 & 1 & -2 \\ 1 & -2 & 1 \end{bmatrix} = \begin{bmatrix} 1 & -2 & 1 \\ -2 & 5 & -4 \\ 1 & -4 & 6 \end{bmatrix}.$$

Exercise 4.6.1

With pivoting:

$$\begin{bmatrix} 0.962 & 0.0149 & 1.000 \\ 0.00137 & 0.859 & 1.000 \end{bmatrix} \rightarrow \begin{bmatrix} 0.962 & 0.0149 & 1.000 \\ 0 & 0.859 & 0.999 \end{bmatrix},$$

where the multiplier used was 0.00142. Solving by back-substitution gives (1.02, 1.16), which is correct to 3 figures.

Without pivoting:

$$\begin{bmatrix} 0.00137 & 0.859 & 1.000 \\ 0.962 & 0.0149 & 1.000 \end{bmatrix} \rightarrow \begin{bmatrix} 0.00137 & 0.859 & 1.000 \\ 0 & -603 & -701 \end{bmatrix},$$

where the multiplier was 702. Now we find $y = 1.16$ as before, but putting this into the first equation,

$$x = \frac{1.000 - 0.859 \times 1.16}{0.00137} \approx \frac{1.000 - 0.996}{0.00137} = \frac{0.004}{0.00137} \approx 2.92$$

The error occurs when the nearly equal numbers 1.000 and 0.996 are subtracted causing the loss of two figures' accuracy.

Miscellaneous Exercises 4

1      $p(0) = \sin 0 \Rightarrow c = 0$, $p\left(\frac{\pi}{2}\right) = \sin \frac{\pi}{2} \Rightarrow a\frac{\pi^2}{4} + b\frac{\pi}{2} = 1$,

$$p(\pi) = \sin \pi \Rightarrow a\pi^2 + b\pi = 0;$$

solving these gives

$$a = -\frac{4^2}{\pi}, \, b = -\frac{4}{\pi}.$$

The required quadratic is then

$$p(x) = \frac{4}{\pi} x\left(1 - \frac{x}{\pi}\right).$$

$$\sin \frac{\pi}{3} \approx p\left(\frac{\pi}{3}\right) = \frac{4}{3}\left(1 - \frac{1}{3}\right) = \frac{8}{9}.$$

$$\left(\text{Actual value } \frac{\sqrt{3}}{2} \approx 0.866. \right)$$

2      $x_1 - x_4 = 100$, $x_2 - x_1 = 200$, $x_2 - x_3 = 100$, $x_3 - x_4 = 200$.

With $x_4 = 100$,

$$x_3 = 300, \, x_2 = 400, \, x_1 = 200.$$

3      $i_1 = 1, \, i_2 = 2, \, i_3 = 1$.

4      $$X^{\mathrm{T}}X = \begin{bmatrix} n & \sum_{i=1}^{n} x_i \\ \sum_{i=1}^{n} x_i & \sum_{i=1}^{n} x_i^2 \end{bmatrix} = \begin{bmatrix} 5 & 15 \\ 15 & 55 \end{bmatrix}.$$

The equations for a, b are then

$$5a_0 + 15a_1 = 17$$

$$15a_0 + 55a_1 = 63,$$

which have the solution $a_0 = -0.2$, $a_1 = 1.2$ so that the least squares regression line is given by $f(x) = 1.2x - 0.2$,

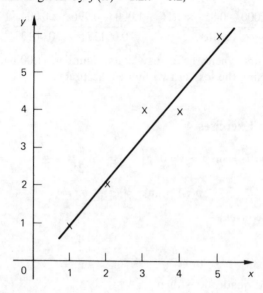

5 Scaling and proceeding as suggested:

$$\begin{bmatrix} 0.00159 & 1.000 & 1.16 \\ 1.000 & 0.0155 & 1.04 \end{bmatrix} \rightarrow \begin{bmatrix} 1.000 & 0.0155 & 1.04 \\ 0.00159 & 1.000 & 1.16 \end{bmatrix}$$

$$\rightarrow \begin{bmatrix} 1.000 & 0.0155 & 1.04 \\ 0 & 1.000 & 1.16 \end{bmatrix}.$$

Thus $y = 1.16$, $x = 1.04 - 0.0155 \times 1.16 \approx 1.02$.

5 VECTORS

5.1 INTRODUCTION

Up to now we have used real numbers to represent the magnitude of physical quantities, such as mass and length, which are unrelated to any direction in space. We shall call such quantities *scalars*: they obey the ordinary rules of algebra. However, there are many physical quantities, such as velocity and force, which are only specified completely when a direction is given as well as a magnitude. We call such quantities *vectors* and we shall develop an algebra for their manipulation. In many applications we shall find that this leads to very elegant and concise solutions. However, when we have to apply the results to a practical situation, we usually find it necessary to describe the vectors in terms of their coordinates. For this reason, as well as to ease the derivation of algebraic rules for vectors, we start with a brief study of coordinate systems.

5.2 COORDINATE SYSTEMS

In order to measure the lengths and directions of lines, we need some kind of reference system.

In two-dimensional space, we set up mutually perpendicular axes, intersecting at an origin, O. We refer to these axes as rectangular cartesian axes and designate the positive directions of the axes as Ox (the x axis) and Oy (the y axis). The x and y coordinates of a point P are defined by the perpendicular distances, x' and y', of P from Oy and Ox respectively, as shown in Figure 5.1. We say that P has coordinates (x', y'). We note that the length of OP can be obtained, using Pythagoras' theorem, as $\sqrt{x'^2 + y'^2}$.

In three dimensions, we add a third axis perpendicular to both Ox and Oy. Its positive direction, Oz, is chosen in the following way. Hold your right thumb and first finger in the plane of your right hand and make the second finger perpendicular to both. With your thumb pointing along Ox

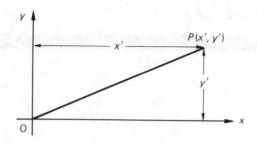

Fig. 5.1

and your first finger along Oy, choose Oz in the direction your second finger points. A set of three vectors with this property is called a *right-handed triad*, and $Oxyz$ is called a *right-handed cartesian coordinate system*. The point P with coordinates (x', y', z') is illustrated in a perspective drawing in Figure 5.2. Here Oyz is in the plane of the paper, while Ox is towards the reader. x', y', z' are, respectively, the perpendicular distances of P from the planes Oyz, Ozx and Oxy. We shall refer to these planes as the yz plane, zx plane and xy plane, respectively; their equations are $x = 0$, $y = 0$ and $z = 0$. The equation $x = x'$ represents a plane parallel to the yz plane and at a distance x' from it.

The length of OP is obtained by Pythagoras' theorem applied to triangle $OP'P$ as $\sqrt{x'^2 + y'^2 + z'^2}$.

The distance between the point P and another point Q with coordinates (x'', y'', z'') may be obtained by moving the origin O to be at P without changing the direction of the axes, as shown in Figure 5.3. If the coordinates of Q measured from this new origin are (x, y, z), then the x coordinate of Q is $x' + x = x''$, giving $x = x'' - x'$. Similarly, $y = y'' - y'$ and $z = z'' - z'$. From

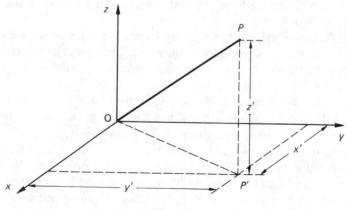

Fig. 5.2

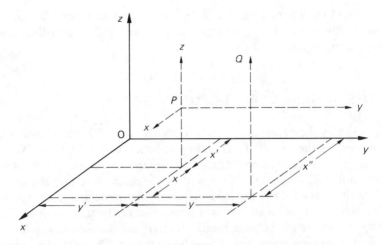

Fig. 5.3

our earlier result, we obtain the length of PQ as

$$\sqrt{(x''-x')^2+(y''-y')^2+(z''-z')^2}.$$

Example 5.2.1

Find all the points which are a distance 5 from the origin and a perpendicular distance $2\sqrt{2}$ from the xy and zx planes.

Suppose that one of the points has coordinates (x, y, z). Then $y = \pm 2\sqrt{2}$, $z = \pm 2\sqrt{2}$, and since $OP = 5$, we must have $x^2 + y^2 + z^2 = 25$. Hence, $x^2 = 25 - 16 = 9$, giving $x = \pm 3$.

The complete solution thus consists of the eight points $(3, 2\sqrt{2}, 2\sqrt{2})$, $(3, 2\sqrt{2}, -2\sqrt{2})$, $(3, -2\sqrt{2}, 2\sqrt{2})$, $(3, -2\sqrt{2}, -2\sqrt{2})$, $(-3, 2\sqrt{2}, 2\sqrt{2})$, $(-3, 2\sqrt{2}, -2\sqrt{2})$, $(-3, -2\sqrt{2}, 2\sqrt{2})$ and $(-3, -2\sqrt{2}, -2\sqrt{2})$.

EXERCISES 5.2.1

1 Find the distance between the points $(5, -5, 2)$ and $(-3, -1, 1)$.

2 Find all the points which are a perpendicular distance $\dfrac{1}{\sqrt{2}}$ from the x, y and z axes.

3 A point O' has coordinates $(1, 1, -1)$ with respect to $Oxyz$. New axes $O'x$, $O'y$ and $O'z$ are set up at O' such that they are parallel to Ox, Oy

and Oz. Find the coordinates of O with respect to O'xyz. If a point P has coordinates $(-1, 2, 0)$ with respect to O'xyz, find its coordinates with respect to Oxyz.

5.3 THE ALGEBRA OF VECTORS

A vector may be represented by a directed line, $\overrightarrow{PQ}$. Its magnitude is represented by the length of PQ and its direction is from P to Q. We shall use a lower case letter in bold type—for example, **r**—as standard notation for a vector. The magnitude of **r** is written as $|\mathbf{r}|$, or more simply as r.

We say that $\mathbf{r} = \mathbf{r}'$ if **r** and **r**' have the same magnitude and direction. Note the implication of this: a vector **r** can be represented by many different lines, so long as they are all the same length and have the same direction. A vector so defined is quite independent of coordinate systems. Nevertheless, it makes the development of vectors easier if we describe them in terms of a cartesian coordinate system, Oxyz.

Now suppose that $\mathbf{r} = \overrightarrow{PQ}$, where the points P and Q have coordinates (x', y', z') and (x'', y'', z''), respectively. If $x'' - x' = x, y'' - y' = y$ and $z'' - z' = z$, then we call x, y and z the *components* of the vector **r** with respect to Oxyz, and we write

$$\overrightarrow{PQ} = \mathbf{r} = (x, y, z)$$

From Section 5.2, the magnitude of **r** is $r = \sqrt{x^2 + y^2 + z^2}$.

With respect to a particular coordinate system Oxyz, the vector joining the origin to a point P, $\overrightarrow{OP} = \mathbf{r} = (x, y, z)$, say, is called the *position vector* of P.

Addition of Vectors

The sum of two vectors $\mathbf{r} = (x, y, z)$ and $\mathbf{r}' = (x', y', z')$ is defined by

$$\mathbf{r} + \mathbf{r}' = (x + x', y + y', z + z')$$

This sum is another vector, since it is a three-component number, just as **r** and **r**' are. we have chosen to define the sum algebraically for simplicity; the geometric equivalent can be derived as follows.

Let **r** be represented by $\overrightarrow{OX}$ and **r**' by $\overrightarrow{XX'}$, as shown in Figure 5.4. Then from Section 5.2, the coordinates $(x + x', y + y', z + z')$ appearing in $\mathbf{r} + \mathbf{r}'$ correspond to the point X', so that $\mathbf{r} + \mathbf{r}'$ is represented by $\overrightarrow{OX'}$.

The geometric interpretation of the sum of **r** and **r**' is simply that it is the vector represented by the third side of the triangle whose two other sides represent the vectors **r** and **r**'. This is called the *triangle law* or, if we include $\overrightarrow{OP}$ and $\overrightarrow{PX'}$, which represent **r**' and **r** respectively, the *parallelogram* law. We note that it is quite independent of any coordinate frame.

134

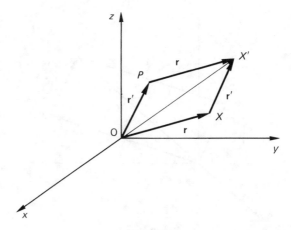

Fig. 5.4

Addition of vectors has the following properties:

(1) *Commutativity*: $\mathbf{r} + \mathbf{r}' = \mathbf{r}' + \mathbf{r}$. This is easily verified from the definition:

$$\mathbf{r} + \mathbf{r}' = (x + x', \, y + y', \, z + z')$$
$$= (x' + x, \, y' + y, \, z' + z)$$
$$= \mathbf{r}' + \mathbf{r}.$$

(2) *Associativity*: $(\mathbf{r} + \mathbf{r}') + \mathbf{r}'' = \mathbf{r} + (\mathbf{r}' + \mathbf{r}'')$. This can also be verified from the definition.

Multiplication of a Vector by a Scalar

If s is a scalar, then the vector $s\mathbf{r}$ is a vector whose magnitude is s times that of $\mathbf{r}$, whose direction is the same as that of $\mathbf{r}$, but whose *sense* is the same as that of $\mathbf{r}$ if s is positive and opposite to that of $\mathbf{r}$ if s is negative. When we multiply $\mathbf{r}$ by s, we multiply its components by s; we thus have

$$s\mathbf{r} = (sx, \, sy, \, sz)$$

If we take the particular value -1 for s, we obtain the vector $(-1)\mathbf{r}$, which we write as $-\mathbf{r}$. It is a vector with the same magnitude and direction as $\mathbf{r}$, but with the opposite sense. In component form this gives

$$-\mathbf{r} = (-x, \, -y, \, -z)$$

We can define the subtraction of a vector $\mathbf{r}$ from a vector $\mathbf{r}'$ as follows:

$$\mathbf{r}' - \mathbf{r} = \mathbf{r}' + (-\mathbf{r})$$

This allows us to obtain the vector $\overrightarrow{XX'}$ in terms of the position vectors $\mathbf{r}$ and $\mathbf{r}'$ of X and X'. In Figure 5.5 we have completed the parallelogram by

135

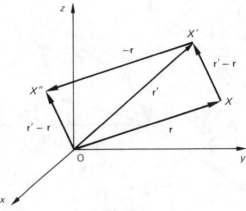

Fig. 5.5

including the vectors $\overrightarrow{X'X''} = -\mathbf{r}$ and $\overrightarrow{OX''} = \overrightarrow{XX'}$, which is the vector we require. Using the triangle law,

$$\overrightarrow{XX'} = \overrightarrow{OX''} = \mathbf{r'} + (-\mathbf{r}) = \mathbf{r'} - \mathbf{r} \tag{1}$$

This is a useful result to remember.

Now

$$\mathbf{r} + (-\mathbf{r}) = (x - x,\ y - y,\ z - z)$$
$$= (0,\ 0,\ 0)$$

We call this the *zero vector* and write it as **0**. It has zero magnitude and no particular direction and is really just an algebraic convenience.

Example 5.3.1

The points A, B, C, D form a parallelogram. A, B, C have position vectors **a**, **b**, **c**. Find the position vector of D.

Let **d** be the position vector of D. Then $\overrightarrow{CD} = \overrightarrow{BA}$, since opposite sides of a parallelogram are equal and parallel. Putting position vectors into this equation and using the result (1) gives

$$\mathbf{d} - \mathbf{c} = \mathbf{a} - \mathbf{b}$$

Adding **c** to both sides, we obtain

$$\mathbf{d} - \mathbf{c} + \mathbf{c} = \mathbf{a} - \mathbf{b} + \mathbf{c}$$

So

$$\mathbf{d} + \mathbf{0} = \mathbf{a} - \mathbf{b} + \mathbf{c}$$

Since the left-hand side is just **d**, the required position vector of D is $\mathbf{a} - \mathbf{b} + \mathbf{c}$.

EXERCISES 5.3.1

1 Let $\mathbf{a} = (1, 2, 3)$ and $\mathbf{b} = (-3, 1, -1)$. Find $\mathbf{a} + 3\mathbf{b}$, $\mathbf{b} - 2\mathbf{a}$ and $|\tfrac{1}{2}\mathbf{a} + \mathbf{b}|$.

2 Points A, B, C, D, which are not necessarily in the same plane, are joined to form a quadrilateral. P, Q, R and S are the midpoints of AB, BC, CD and DA, respectively. Use vector methods to show that $\overrightarrow{PQ} = \tfrac{1}{2}\overrightarrow{AC}$. Deduce that $PQRS$ is a parallelogram.

3 Points A, B, C, D form a parallelogram and M is the midpoint of AB. The lines DM and AC intersect at X. The position vectors of B and D with respect to A are $\mathbf{a}$ and $\mathbf{b}$, respectively, and $\overrightarrow{AX} = \lambda\overrightarrow{AC}$, $\overrightarrow{DX} = \mu\overrightarrow{DM}$. Express

(i) $\overrightarrow{AX}$ in terms of $\mathbf{a}$, $\mathbf{b}$ and λ, (ii) $\overrightarrow{AX}$ in terms of $\mathbf{a}$, $\mathbf{b}$ and μ, and (iii) deduce that DM trisects AC.

5.4 UNIT VECTORS AND DIRECTION COSINES

Unit Vectors

A vector of unit magnitude is called a *unit vector*. If $\mathbf{r}$ is any vector, the unit vector in the direction of $\mathbf{r}$ is written as $\hat{\mathbf{r}}$.

Example 5.4.1

Find a unit vector in the direction of $\mathbf{s} = (2, 3, 4)$.
We have $s^2 = 4 + 9 + 16 = 29$. Hence,

$$\hat{\mathbf{s}} = \mathbf{s}/s = \left(\frac{2}{\sqrt{29}}, \frac{3}{\sqrt{29}}, \frac{4}{\sqrt{29}} \right)$$

The particular unit vectors in the directions of the coordinate axes, Ox, Oy and Oz are denoted by $\mathbf{i}$, $\mathbf{j}$ and $\mathbf{k}$, respectively. In component form these are

$$\mathbf{i} = (1, 0, 0), \mathbf{j} = (0, 1, 0) \text{ and } \mathbf{k} = (0, 0, 1)$$

These unit vectors provide another way of expressing vectors. For example,

$$\mathbf{r} = (x, y, z)$$
$$= (x, 0, 0) + (0, y, 0) + (0, 0, z)$$
$$= x\mathbf{i} + y\mathbf{j} + z\mathbf{k}$$

Direction Cosines

Let the angles between the vector $\mathbf{r} = (x, y, z)$ and the vectors $\mathbf{i}$, $\mathbf{j}$ and $\mathbf{k}$ be α, β and γ, respectively. Then, as we can see from Figure 5.6, $\cos \alpha = x/r$, $\cos \beta = y/r$ and $\cos \gamma = z/r$. Thus,

$$\hat{\mathbf{r}} = \frac{\mathbf{r}}{r} = \frac{x\mathbf{i} + y\mathbf{j} + z\mathbf{k}}{r}$$

$$= \mathbf{i} \cos \alpha + \mathbf{j} \cos \beta + \mathbf{k} \cos \gamma$$

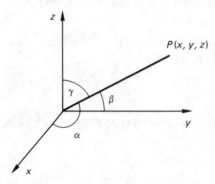

Fig. 5.6

The numbers $l = \cos \alpha$, $m = \cos \beta$ and $n = \cos \gamma$ are called the *direction cosines* of $\mathbf{r}$ and they enable us to write

$$\hat{\mathbf{r}} = l\mathbf{i} + m\mathbf{j} + n\mathbf{k}$$

We note that these numbers are not independent, since

$$l^2 + m^2 + n^2 = \left(\frac{x}{r}\right)^2 + \left(\frac{y}{r}\right)^2 + \left(\frac{z}{r}\right)^2$$

$$= \frac{r^2}{r^2} = 1$$

The direction cosines of a line L not going through the origin are defined to be equal to those of the line parallel to L through the origin.

Example 5.4.2

Find the direction cosines of a line inclined at an angle of $\pi/4$ to the x axis and $\pi/3$ to the y axis, and, hence, the unit vector in the direction of the line.

$$l = \cos \frac{\pi}{4} = \frac{1}{\sqrt{2}}, \quad m = \cos \frac{\pi}{3} = \frac{1}{2} \text{ and, since } l^2 + m^2 + n^2 = 1, \text{ we obtain}$$

$n = \sqrt{1 - 1/2 - 1/4} = \pm\dfrac{1}{2}$. There are two possible solutions, $(l, m, n) = \left(\dfrac{1}{\sqrt{2}}, \dfrac{1}{2}, \dfrac{1}{2}\right)$ with L above the xy plane and $(l, m, n) = \left(\dfrac{1}{\sqrt{2}}, \dfrac{1}{2}, -\dfrac{1}{2}\right)$ with L below.

Angle between Two Lines

Let the two lines L, L' through the origin have direction cosines l, m, n and l', m', n', respectively, and have an angle θ between them. The points P and P' at unit distance from the origin on these lines have coordinates (l, m, n) and (l', m', n'), so that the length of PP' is given by

$$(PP')^2 = (l' - l)^2 + (m' - m)^2 + (n' - n)^2$$

The cosine rule applied to the triangle OPP' then gives

$$\cos\theta = \frac{OP^2 + (OP')^2 - (PP')^2}{2 \cdot OP \cdot OP'}$$

$$= 1 - \frac{1}{2}(PP')^2$$

$$= 1 - ((l'^2 + m'^2 + n'^2) - 2(ll' + mm' + nn') + (l^2 + m^2 + n^2))/2$$

$$= ll' + mm' + nn'$$

since

$$l^2 + m^2 + n^2 = l'^2 + m'^2 + n'^2 = 1$$

A particularly important result is that the two lines are perpendicular if their direction cosines satisfy

$$ll' + mm' + nn' = 0$$

EXERCISES 5.4.1

1 Two lines L_1 and L_2, which pass through the origin, have direction cosines $\sqrt{3}/2$, $\frac{1}{2}$, 0 and 0, 0, -1, respectively. Show that L_1 and L_2 are perpendicular, and find the direction cosines of a third line, L_3, which is perpendicular to both L_1 and L_2.

2 The origin of cartesian coordinates is at the centre of a cube whose edges are parallel to the axes. Find the angle between any two of the diagonals of the cube.

5.5 SCALAR PRODUCTS

The *scalar or dot product*, $\mathbf{r} \cdot \mathbf{r}'$, of two vectors $\mathbf{r} = (x, y, z)$ and $\mathbf{r}' = (x', y', z')$ is defined by $\mathbf{r} \cdot \mathbf{r}' = xx' + yy' + zz'$. Note that this is a scalar quantity, despite being formed as a product of two vectors. When $\mathbf{r}' = \mathbf{r}$, we write

$$\mathbf{r} \cdot \mathbf{r} = r^2$$
$$= x^2 + y^2 + z^2$$
$$= r^2$$

Properties

(1)
$$\mathbf{r} \cdot \mathbf{r}' = xx' + yy' + zz'$$
$$= x'x + y'y + z'z$$
$$= \mathbf{r}' \cdot \mathbf{r}$$

Thus, scalar multiplication is commutative.

(2) Let s be a scalar. Then
$$\mathbf{r} \cdot (s\mathbf{r}') = (x, y, z) \cdot (sx', sy', sz')$$
$$= sxx' + syy' + szz'$$
$$= s(xx' + yy' + zz')$$
$$= s(\mathbf{r} \cdot \mathbf{r}')$$

Thus, any scalar can simply be moved to the front, leaving the vectors to form the scalar product.

(3) Let $\mathbf{r}'' = (x'', y'', z'')$. Then
$$\mathbf{r} \cdot (\mathbf{r}' + \mathbf{r}'') = x(x' + x'') + y(y' + y'') + z(z' + z'')$$
$$= xx' + yy' + zz' + xx'' + yy'' + zz''$$
$$= \mathbf{r} \cdot \mathbf{r}' + \mathbf{r} \cdot \mathbf{r}''$$

This says that scalar multiplication is distributive over vector addition.

Geometric Interpretation

The direction cosines of $\mathbf{r} = (x, y, z)$ and $\mathbf{r}' = (x', y', z')$ are x/r, y/r, z/r and x'/r', y'/r', z'/r', respectively. The angle, θ, between them is given by

$$\cos \theta = \frac{x}{r} \cdot \frac{x'}{r'} + \frac{y}{r} \cdot \frac{y'}{r'} + \frac{z}{r} \cdot \frac{z'}{r'} = \frac{\mathbf{r} \cdot \mathbf{r}'}{rr'}$$

so that we have

$$\mathbf{r} \cdot \mathbf{r}' = rr' \cos \theta$$

This says that the scalar product of two vectors is the product of their magnitudes times the cosine of the angle between them. This is clearly independent of the coordinate system used.

Since $\cos \theta$ is zero if and only if θ is an odd multiple of $\pi/2$, two non-zero vectors are perpendicular if and only if their scalar product is zero. In particular, this means that the unit vectors $\mathbf{i}, \mathbf{j}$ and $\mathbf{k}$ have the property that

$$\mathbf{i} \cdot \mathbf{j} = \mathbf{j} \cdot \mathbf{k} = \mathbf{k} \cdot \mathbf{i} = 0$$

Because they are unit vectors, they also satisfy

$$\mathbf{i}^2 = \mathbf{j}^2 = \mathbf{k}^2 = 1$$

We can see that the rules are consistent by forming the scalar multiple of the vectors expressed in terms of $\mathbf{i}, \mathbf{j}$ and $\mathbf{k}$:

$$\mathbf{r} \cdot \mathbf{r}' = (x\mathbf{i} + y\mathbf{j} + z\mathbf{k}) \cdot (x'\mathbf{i} + y'\mathbf{j} + z'\mathbf{k})$$
$$= xx'\mathbf{i}^2 + yy'\mathbf{j}^2 + zz'\mathbf{k}^2 + xy'\mathbf{i} \cdot \mathbf{j} + xz'\mathbf{i} \cdot \mathbf{k}$$
$$+ yx'\mathbf{j} \cdot \mathbf{i} + yz'\mathbf{j} \cdot \mathbf{k} + zx'\mathbf{k} \cdot \mathbf{i} + zy'\mathbf{k} \cdot \mathbf{j}$$
$$= xx' + yy' + zz'$$

since all the other products are zero.

Example 5.5.1

Find the angle between the vectors $\mathbf{r} = \mathbf{i} + 2\mathbf{j} - 3\mathbf{k}$ and $\mathbf{r}' = -\mathbf{i} - 2\mathbf{k}$.

Let the required angle be θ. Then

$$\cos \theta = \frac{\mathbf{r} \cdot \mathbf{r}'}{rr'}$$

$$= \frac{-1 + 0 + 6}{\sqrt{1 + 4 + 9}\sqrt{1 + 4}}$$

$$= \frac{5}{\sqrt{14}\sqrt{5}}$$

giving

$$\theta = \cos^{-1}\sqrt{\frac{5}{14}}$$

EXERCISES 5.5.1

1 Find the angle between the vectors $(0, -1, 1)$ and $(3, 4, 5)$.
2 Show that the vectors $(1, 8, 2)$ and $(2\lambda^2, -\lambda, 4)$ are perpendicular if and only if $\lambda = 2$.
3 In a triangle ABC the perpendiculars from A, B to BC, AC, respectively, intersect at O. Using the results of this section, show that if $\overrightarrow{OA} = \mathbf{a}$, $\overrightarrow{OB} = \mathbf{b}$, $\overrightarrow{OC} = \mathbf{c}$, then

$$\mathbf{a} \cdot (\mathbf{b} - \mathbf{c}) = \mathbf{b} \cdot (\mathbf{a} - \mathbf{c}) = 0$$

Deduce that $\mathbf{c} \cdot (\mathbf{a} - \mathbf{b}) = 0$ and interpret this result.

5.6 VECTOR PRODUCTS

The *vector or cross product* of $\mathbf{r} = (x, y, z)$ and $\mathbf{r}' = (x', y', z')$ is written $\mathbf{r} \times \mathbf{r}'$ and defined by

$$\mathbf{r} \times \mathbf{r}' = (yz' - y'z, zx' - z'x, xy' - x'y)$$

Unlike the scalar product, this is a vector, whose direction we shall shortly investigate. In order to make the definition more memorable, we shall make use of determinant notation (see Section 4.2). Although the elements of determinants are normally scalars, we shall put the vectors $\mathbf{i}$, $\mathbf{j}$ and $\mathbf{k}$ in the first row and assume that the normal rules for the expansion still apply. Thus, we may write the vector product as

$$\mathbf{r} \times \mathbf{r}' = \begin{vmatrix} \mathbf{i} & \mathbf{j} & \mathbf{k} \\ x & y & z \\ x' & y' & z' \end{vmatrix}$$

Expanding formally by the first row, we obtain the previous definition.

Before examining the properties further, we introduce the *triple scalar product*, $\mathbf{r} \cdot (\mathbf{r}' \times \mathbf{r}'')$. Using our usual notation for the components of $\mathbf{r}$, $\mathbf{r}'$ and $\mathbf{r}''$, we have

$$\mathbf{r} \cdot (\mathbf{r}' \times \mathbf{r}'') = (x, y, z) \cdot (y'z'' - z'y'', z'x'' - x'z'', x'y'' - y'x'')$$

$$= x(y'z'' - z'y'') + y(z'x'' - x'z'') + z(x'y'' - y'x'')$$

$$= \begin{vmatrix} x & y & z \\ x' & y' & z' \\ x'' & y'' & z'' \end{vmatrix}$$

We shall return to the triple scalar product later, but now we use it to find out what the vector product means.

Geometric Interpretation of the Vector Product

Using the last result, we see that both $\mathbf{r} \cdot (\mathbf{r} \times \mathbf{r}')$ and $\mathbf{r}' \cdot (\mathbf{r} \times \mathbf{r}')$ are zero, since the determinant form of each contains two identical rows. This shows that $\mathbf{r} \times \mathbf{r}'$ is a vector perpendicular to both $\mathbf{r}$ and $\mathbf{r}'$. We now investigate the magnitude of $\mathbf{r} \times \mathbf{r}'$. We have

$$\mathbf{r} \times \mathbf{r}' = \begin{vmatrix} \mathbf{i} & \mathbf{j} & \mathbf{k} \\ x & y & z \\ x' & y' & z' \end{vmatrix}$$

Dividing by r and r', we obtain

$$\left(\frac{\mathbf{r}}{r}\right) \times \left(\frac{\mathbf{r}'}{r'}\right) = \begin{vmatrix} \mathbf{i} & \mathbf{j} & \mathbf{k} \\ \dfrac{x}{r} & \dfrac{y}{r} & \dfrac{z}{r} \\ \dfrac{x'}{r'} & \dfrac{y'}{r'} & \dfrac{z'}{r'} \end{vmatrix}$$

$$= \begin{vmatrix} \mathbf{i} & \mathbf{j} & \mathbf{k} \\ l & m & n \\ l' & m' & n' \end{vmatrix}$$

where l, m, n and l', m', n' are the direction cosines of $\mathbf{r}$ and $\mathbf{r}'$, respectively. The square of the magnitude of this vector is

$$(mn' - m'n)^2 + (nl' - n'l)^2 + (lm' - l'm)^2$$
$$= (l^2 + m^2 + n^2)(l'^2 + m'^2 + n'^2) - (ll' + mm' + nn')^2$$
$$= 1 - \cos^2 \theta$$
$$= \sin^2 \theta$$

where θ is the angle between the two vectors.
  Putting these results together, we have

$$\frac{\mathbf{r}}{r} \times \frac{\mathbf{r}'}{r'} = \mathbf{n} \sin \theta$$

where $\mathbf{n}$ is a unit vector perpendicular to both $\mathbf{r}$ and $\mathbf{r}'$, which gives

$$\mathbf{r} \times \mathbf{r}' = \mathbf{n}rr' \sin \theta$$

The only remaining problem is to find the sense of $\mathbf{n}$. This we do by using the definition to evaluate $\mathbf{r} \times \mathbf{r}'$, where $\mathbf{r} = (1, 0, 0), \mathbf{r}' = (x, y, 0)$. This is a perfectly general choice if $xy \neq 0$, and gives

$$\mathbf{r}' \times \mathbf{r}' = \begin{vmatrix} \mathbf{i} & \mathbf{j} & \mathbf{k} \\ 1 & 0 & 0 \\ x & y & 0 \end{vmatrix}$$

$$= y\mathbf{k}$$

If $y > 0$ we have the case illustrated in Fig. 5.9, where the product is in the sense of increasing x, while if $y < 0$ the product is in the opposite sense. Both cases correspond to a right-handed triad. Thus the direction of a vector product is such that the two vectors and their product form a right-handed triad.

Vector Product Properties

(1) The vector product is not commutative. in fact,

$$\mathbf{r} \times \mathbf{r}' = -\mathbf{r}' \times \mathbf{r}$$

This result is easily verified, using determinant properties, as follows:

$$\mathbf{r} \times \mathbf{r}' = \begin{vmatrix} \mathbf{i} & \mathbf{j} & \mathbf{k} \\ x & y & z \\ x' & y' & z' \end{vmatrix} = -\begin{vmatrix} \mathbf{i} & \mathbf{j} & \mathbf{k} \\ x' & y' & z' \\ x & y & z \end{vmatrix} = -\mathbf{r}' \times \mathbf{r}$$

(2) The vector product is distributive over vector addition, that is,

$$\mathbf{r} \times (\mathbf{r}' + \mathbf{r}'') = \mathbf{r} \times \mathbf{r}' + \mathbf{r} \times \mathbf{r}''$$

Using determinant properties again, we have

$$\mathbf{r} \times (\mathbf{r}' + \mathbf{r}'') = \begin{vmatrix} \mathbf{i} & \mathbf{j} & \mathbf{k} \\ x & y & z \\ x' + x'' & y' + y'' & z' + z'' \end{vmatrix}$$

Expanding by the last row and rearranging gives

$$\mathbf{r} \times (\mathbf{r}' + \mathbf{r}'') = \begin{vmatrix} \mathbf{i} & \mathbf{j} & \mathbf{k} \\ x & y & z \\ x' & y' & z' \end{vmatrix} + \begin{vmatrix} \mathbf{i} & \mathbf{j} & \mathbf{k} \\ x & y & z \\ x'' & y'' & z'' \end{vmatrix}$$

$$= \mathbf{r} \times \mathbf{r}' + \mathbf{r} \times \mathbf{r}''$$

(3) The following relationships should be verified from the definitions:

$$\mathbf{j} \times \mathbf{k} = -\mathbf{k} \times \mathbf{j} = \mathbf{i} \qquad \mathbf{i} \times \mathbf{j} = -\mathbf{j} \times \mathbf{i} = \mathbf{k}$$

$$\mathbf{k} \times \mathbf{i} = -\mathbf{i} \times \mathbf{k} = \mathbf{j} \qquad \mathbf{i} \times \mathbf{i} = \mathbf{j} \times \mathbf{j} = \mathbf{k} \times \mathbf{k} = 0$$

Examples 5.6.1

1. Find the most general form of vector **r** satisfying the equation $\mathbf{r} \times (1, 1, 1) = (2, -4, 2)$. Find that value of **r** which satisfies $\mathbf{r} \cdot (1, 1, 1) = 0$.

   With the usual notation for the components of **r**, we have

   $$\mathbf{r} \times (1, 1, 1) = (y - z, z - x, x - y)$$

   This equals $(2, -4, 2)$ if $y - z = 2$, $z - x = -4$ and $x - y = 2$. We cannot find x, y and z uniquely, but if we let $x = t$, say, then $y = t - 2$, $z = t - 4$, and so we may write the required form of **r** as $(t, t - 2, t - 4)$.

   To satisfy $\mathbf{r} \cdot (1, 1, 1) = 0$, we must have $t + (t - 2) + (t - 4) = 0$, or $t = 2$, giving the particular **r** required as $(2, 0, -2)$.

2. Show that the area of the parallelogram $PQRS$, where $\overrightarrow{PQ} = \mathbf{r}$ and $\overrightarrow{PS} = \mathbf{r}'$ is $|\mathbf{r} \times \mathbf{r}'|$.

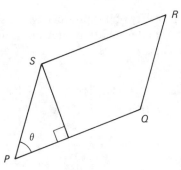

Fig. 5.7

The area of a parallelogram (see Figure 5.7) is obtained by multiplying PQ by the perpendicular distance between PQ and RS. If θ is the angle between PQ and PS, then

$$\text{area } PQRS = PQ \cdot PS \sin \theta$$

$$= rr' \sin \theta$$

but this is just $|\mathbf{r} \times \mathbf{r}'|$.

EXERCISES 5.6.1

1 Find a and b so that $(a, b, 1) \times (2, 1, 5) = (1, 3, -1)$.

2 By constructing a *numerical* example, show that, in general, the associative law for vector products does not hold, that is, find an example to show that

$$\mathbf{a} \times (\mathbf{b} \times \mathbf{c}) \neq (\mathbf{a} \times \mathbf{b}) \times \mathbf{c}$$

3 Find the most general form of vector **u** satisfying

$$\mathbf{u} \times (2\mathbf{i} + \mathbf{j} - \mathbf{k}) = \mathbf{i} \times (2\mathbf{i} + \mathbf{j} - \mathbf{k})$$

4 By using a geometric argument prove that if $\mathbf{a} \times \mathbf{b} = \mathbf{a} - \mathbf{b}$, then $\mathbf{a} = \mathbf{b}$.

5.7 TRIPLE PRODUCTS

We have already seen in Section 5.6 that the triple scalar product, $\mathbf{r} \cdot (\mathbf{r}' \times \mathbf{r}'')$, can be expressed as the determinant

$$\begin{vmatrix} x & y & z \\ x' & y' & z' \\ x'' & y'' & z'' \end{vmatrix}$$

We now investigate a property of this product, which seems remarkable until we examine its geometric interpretation later on. Using our usual notation for the vector components, we have

$$\mathbf{r} \cdot (\mathbf{r}' \times \mathbf{r}'') = \begin{vmatrix} x & y & z \\ x' & y' & z' \\ x'' & y'' & z'' \end{vmatrix}$$

$$= - \begin{vmatrix} x'' & y'' & z'' \\ x' & y' & z' \\ x & y & z \end{vmatrix}$$

$$= \begin{vmatrix} x'' & y'' & z'' \\ x & y & z \\ x' & y' & z' \end{vmatrix}$$

$$= \mathbf{r}'' \cdot (\mathbf{r} \times \mathbf{r}')$$

$$= (\mathbf{r} \times \mathbf{r}') \cdot \mathbf{r}''$$

where we have interchanged rows of the determinant twice and used the fact that scalar multiplication is commutative. The result tells us that we can interchange the $\cdot$ and $\times$ without changing the value of the triple scalar product. It is worth noting that if any two of the vectors appearing in the product are the same then the value is zero, since there are two identical rows in the determinant which defines its value.

Example 5.7.1

A parallelepiped has edges formed by vectors $\mathbf{a}$, $\mathbf{b}$ and $\mathbf{c}$. Show that its volume is $|\mathbf{a} \cdot (\mathbf{b} \times \mathbf{c})|$.

The volume is given by the area of a face times the perpendicular distance to the opposite face (see Figure 5.8, in which $\mathbf{a} = \overrightarrow{OA}$, $\mathbf{b} = \overrightarrow{OB}$ and $\mathbf{c} = \overrightarrow{OC}$). If θ is the angle between $\mathbf{b}$ and $\mathbf{c}$ and ϕ is the angle between $\mathbf{a}$ and the unit vector $\mathbf{n}$ perpendicular to the plane containing $\mathbf{b}$ and $\mathbf{c}$, then the required volume is $bc \sin \theta \cdot a \cos \phi$. Now $\mathbf{b} \times \mathbf{c} = \mathbf{n} bc \sin \theta$.

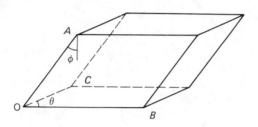

Fig. 5.8

Thus,

$$|\mathbf{a} \cdot (\mathbf{b} \times \mathbf{c})| = |\mathbf{a} \cdot \mathbf{n}| bc \sin \theta$$

$$= a \cos \phi \cdot bc \sin \theta$$

and is, hence, the required volume. The result concerning the interchangeability of $\cdot$ and $\times$ is now simply explained: the two orders correspond to different ways of evaluating the volume of a parallelepiped, which must always be the same.

Triple Vector Product

This is a product of the form $\mathbf{r} \times (\mathbf{r}' \times \mathbf{r}'')$. We show that it satisfies the identity

$$\mathbf{r} \times (\mathbf{r}' \times \mathbf{r}'') = (\mathbf{r} \cdot \mathbf{r}'')\mathbf{r}' - (\mathbf{r} \cdot \mathbf{r}')\mathbf{r}''$$

Choose axes so that Ox lies along $\mathbf{r}$ and Oy lies in the plane of $\mathbf{r}$ and $\mathbf{r}'$, as shown in Figure 5.9. This is a perfectly general choice, provided that $\mathbf{r}$ and $\mathbf{r}'$ are not parallel. With this choice we can take

$$\mathbf{r} = (x, 0, 0), \ \mathbf{r}' = (x', y', 0) \quad \text{and} \quad \mathbf{r}'' = (x'', y'', z'')$$

giving

$$\mathbf{r}' \times \mathbf{r}'' = (y'z'', \ -x'z'', \ x'y'' - y'x'')$$

147

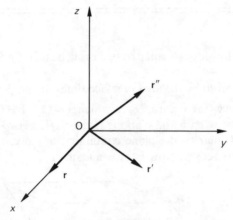

Fig. 5.9

Hence,

$$\mathbf{r} \times (\mathbf{r}' \times \mathbf{r}'') = \begin{vmatrix} \mathbf{i} & \mathbf{j} & \mathbf{k} \\ x & 0 & 0 \\ y'z'' & -x'z'' & x'y'' - y'x'' \end{vmatrix}$$

$$= (0, \; xx''y' - xx'y'', \; -xy'z'')$$

$$= xx''(x', y', 0) - xx'(x'', y'', z'')$$

$$= (\mathbf{r} \cdot \mathbf{r}'')\mathbf{r}' - (\mathbf{r} \cdot \mathbf{r}')\mathbf{r}''$$

Changing the order on the left of this identity and using vector product property (1), we obtain

$$(\mathbf{r}' \times \mathbf{r}'') \times \mathbf{r} = -(\mathbf{r} \cdot \mathbf{r}'')\mathbf{r}' + (\mathbf{r} \cdot \mathbf{r}')\mathbf{r}''$$

Now replace $\mathbf{r}'$ by $\mathbf{r}$, $\mathbf{r}''$ by $\mathbf{r}'$, $\mathbf{r}$ by $\mathbf{r}''$ and rearrange slightly, to obtain

$$(\mathbf{r} \times \mathbf{r}') \times \mathbf{r}'' = (\mathbf{r} \cdot \mathbf{r}'')\mathbf{r}' - (\mathbf{r}' \cdot \mathbf{r}'')\mathbf{r}$$

Examples 5.7.2

1. Show that $(\mathbf{a} \times \mathbf{b}) \cdot (\mathbf{c} \times \mathbf{d}) = (\mathbf{a} \cdot \mathbf{c})(\mathbf{b} \cdot \mathbf{d}) - (\mathbf{a} \cdot \mathbf{d})(\mathbf{b} \cdot \mathbf{c})$.
   Let $\mathbf{c} \times \mathbf{d} = \mathbf{e}$; then the left-hand side becomes

   $$(\mathbf{a} \times \mathbf{b}) \cdot \mathbf{e} = \mathbf{a} \cdot (\mathbf{b} \times \mathbf{e})$$

   $$= \mathbf{a} \cdot (\mathbf{b} \times (\mathbf{c} \times \mathbf{d}))$$

   $$= \mathbf{a} \cdot ((\mathbf{b} \cdot \mathbf{d})\mathbf{c} - (\mathbf{b} \cdot \mathbf{c})\mathbf{d})$$

   $$= (\mathbf{a} \cdot \mathbf{c})(\mathbf{b} \cdot \mathbf{d}) - (\mathbf{a} \cdot \mathbf{d})(\mathbf{b} \cdot \mathbf{c})$$

   as required.

2. Solve the following equations for $\mathbf{x}$ in terms of p, $\mathbf{a}$ and $\mathbf{b}$: $\mathbf{x} \times \mathbf{a} = \mathbf{b}$; $\mathbf{a} \cdot \mathbf{x} = p$.

Taking the vector product of the first equation with $\mathbf{a}$ gives

$$\mathbf{a} \times (\mathbf{x} \times \mathbf{a}) = \mathbf{a} \times \mathbf{b}$$

or

$$(\mathbf{a} \cdot \mathbf{a})\mathbf{x} - (\mathbf{a} \cdot \mathbf{x})\mathbf{a} = \mathbf{a} \times \mathbf{b}$$

This allows us to use the second given equation to substitute p for $\mathbf{a} \cdot \mathbf{x}$ and solve for $\mathbf{x}$, to obtain

$$\mathbf{x} = \frac{\mathbf{a} \times \mathbf{b} + p\mathbf{a}}{a^2}.$$

EXERCISES 5.7.1

1 Check that $\mathbf{a} \cdot \mathbf{b} \times \mathbf{c} = \mathbf{b} \cdot \mathbf{c} \times \mathbf{a} = \mathbf{c} \cdot \mathbf{a} \times \mathbf{b}$. (Note the cyclic order of $\mathbf{a}, \mathbf{b}, \mathbf{c}$.)
2 Justify geometrically the fact that the triple scalar product is zero when two of the vectors are the same.
3 Show that three non-zero vectors, $\mathbf{a}$, $\mathbf{b}$, $\mathbf{c}$, which pass through a common point are coplanar if and only if $\mathbf{a} \cdot \mathbf{b} \times \mathbf{c} = 0$.
4 Without going to components, show that

(i) $\mathbf{a} \times (\mathbf{b} \times (\mathbf{c} \times \mathbf{a})) = (\mathbf{a} \cdot \mathbf{b})(\mathbf{a} \times \mathbf{c})$;
(ii) $(\mathbf{b} \times \mathbf{c}) \times (\mathbf{c} \times \mathbf{a}) = (\mathbf{a} \cdot \mathbf{b} \times \mathbf{c})\mathbf{c}$.

5 Solve for $\mathbf{x}$ the vector equation $\lambda\mathbf{x} + \mathbf{x} \times \mathbf{a} = \mathbf{b}$, where $\mathbf{a}$ and $\mathbf{b}$ are known vectors and λ is a non-zero constant.

5.8 LINES AND PLANES

Vector Equation of a Line

Suppose that we require the equation of a line which passes through the point A, whose position vector is $\mathbf{a}$, and whose direction is given by the vector $\overrightarrow{AB} = \mathbf{b}$. Let $\mathbf{r}$ be the position vector of any point P on the line. $\overrightarrow{AP}$ is just a scalar multiple, t, say, of $\overrightarrow{AB}$, so we can find $\mathbf{r}$, using the triangle law applied to OAP (see Figure 5.10). This gives

$$\mathbf{r} = \overrightarrow{OA} + \overrightarrow{AP}$$

$$= \mathbf{a} + t\mathbf{b}$$

As the value of the parameter t varies, the point P moves along the line. When t is 0 and 1, P coincides with the points A and B, respectively. As t

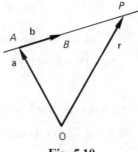

Fig. 5.10

increases from zero, P moves from A along the line in the direction of $\overrightarrow{AB}$. Negative values of t correspond to points on the opposite side of A.

The equation $\mathbf{r} = \mathbf{a} + t\mathbf{b}$ is called the *vector equation* or *parametric equation* of the line. The vector $\mathbf{r}$ is an example of a vector function, since to each value of t there corresponds a value of the vector $\mathbf{r}$. We shall study such functions in Section 5.9.

Example 5.8.1

Find the vector equation of the line through the points $A = (1, 2, 3)$ and $B = (4, 5, 6)$.

We must first find the vector $\overrightarrow{AB}$; we have (referring again to Figure 5.10)

$$\overrightarrow{AB} = \overrightarrow{OB} - \overrightarrow{OA}$$

$$= (4, 5, 6) - (1, 2, 3)$$

$$= (3, 3, 3)$$

Then, from the above derivation, the required equation is

$$\mathbf{r} = (1, 2, 3) + t(3, 3, 3)$$

We note that t is not the only parameter we can use; we could, for example, write the equations as

$$\mathbf{r} = (1, 2, 3) + (3t)(1, 1, 1)$$

$$= (1, 2, 3) + s(1, 1, 1)$$

where we have replaced $3t$ by a new parameter s.

Perpendicular Distance of a Point from a Line

Using the same notation as before, and referring to Figure 5.11, we let N be the point on the line such that $\overrightarrow{ON}$ is perpendicular to the line. Let $\overrightarrow{ON} = \mathbf{n}$.

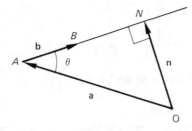

Fig. 5.11

Then, if t' is the value of t at the point N, we have $\mathbf{n} = \mathbf{a} + t'\mathbf{b}$. Since $\vec{ON}$ is perpendicular to $\vec{AB}$, we have

$$0 = \mathbf{b} \cdot \mathbf{n}$$
$$= \mathbf{b} \cdot (\mathbf{a} + t'\mathbf{b})$$
$$= \mathbf{b} \cdot \mathbf{a} + t'\mathbf{b} \cdot \mathbf{b}$$

which gives the value of t' as

$$-\frac{\mathbf{a} \cdot \mathbf{b}}{\mathbf{b} \cdot \mathbf{b}}$$

We now have

$$\mathbf{n} = \mathbf{a} - \frac{\mathbf{a} \cdot \mathbf{b}}{\mathbf{b} \cdot \mathbf{b}} \mathbf{b}$$
$$= \frac{(\mathbf{b} \cdot \mathbf{b})\mathbf{a} - (\mathbf{a} \cdot \mathbf{b})\mathbf{b}}{\mathbf{b} \cdot \mathbf{b}}$$
$$= \frac{(\mathbf{b} \times \mathbf{a}) \times \mathbf{b}}{\mathbf{b} \cdot \mathbf{b}}$$

Since $\mathbf{b} \times \mathbf{a}$ is perpendicular to $\mathbf{b}$, we have

$$ON = |\mathbf{n}| = |\mathbf{b} \times \mathbf{a}| \frac{|\mathbf{b}|}{b^2} = \frac{|\mathbf{b} \times \mathbf{a}|}{b}$$

If θ is the angle between $\vec{AO}$ and $\vec{AN}$, then $ON = OA \sin \theta$, and this last result becomes $ON = ba \sin \theta / b = a \sin \theta$, which is obvious from the triangle OAN. We also note that

$$\vec{NA} = \vec{OA} - \vec{ON}$$
$$= \mathbf{a} - (\mathbf{a} + t'\mathbf{b})$$
$$= \frac{\mathbf{a} \cdot \mathbf{b}}{\mathbf{b} \cdot \mathbf{b}} \mathbf{b}$$

The length of this vector is

$$\frac{|\mathbf{a}\cdot\mathbf{b}||\mathbf{b}|}{b^2} = \frac{ab\cos\theta}{b^2}b = a\cos\theta$$

$\overrightarrow{AN}$ is called the *orthogonal projection of* $\mathbf{a}$ *on* $\mathbf{b}$.

Example 5.8.2

Find the length of the perpendicular from the origin to the line through $A = (1, 2, 3)$ and $B = (4, 5, 6)$.

We found in Example 5.8.1 that the equation of the line was

$$\mathbf{r} = (1, 2, 3) + s(1, 1, 1)$$

Suppose that, at the foot of the perpendicular, N, $s = s'$, then

$$\mathbf{n} = (1, 2, 3) + s'(1, 1, 1)$$

so that

$$0 = \overrightarrow{AB}\cdot\mathbf{n}$$

$$= (3, 3, 3)\cdot(1, 2, 3) + s'(3, 3, 3)\cdot(1, 1, 1)$$

$$= 18 + 9s'$$

so that $s' = -2$ and

$$\mathbf{n} = (1, 2, 3) - 2(1, 1, 1)$$

$$= (-1, 0, 2)$$

giving $n = \sqrt{5}$.

Equation of a Plane

We construct the equation of the plane perpendicular to the vector $\mathbf{n}$ and passing through the point A with position vector $\mathbf{a}$. Let $\mathbf{r}$ be the position vector of any point P in the plane. Then the vector $\mathbf{r} - \mathbf{a}$ lies in the plane and is therefore perpendicular to $\mathbf{n}$. The condition for this is

$$(\mathbf{r} - \mathbf{a})\cdot\mathbf{n} = 0$$

which gives the equation of the plane. We can also write it in the form

$$\mathbf{r}\cdot\mathbf{n} = m$$

which is obtained from the previous form by rearranging and putting $\mathbf{a}\cdot\mathbf{n} = m$.

Example 5.8.3

Find the coordinates of the point P of intersection of the line $\mathbf{r} = \mathbf{a} + t\mathbf{b}$ with the plane $\mathbf{r} \cdot \mathbf{n} = m$.

At the point P, the equations of the line and point must be satisfied simultaneously. Thus,

$$m = (\mathbf{a} + t\mathbf{b}) \cdot \mathbf{n}$$
$$= \mathbf{a} \cdot \mathbf{n} + t\mathbf{b} \cdot \mathbf{n}$$

If $\mathbf{b} \cdot \mathbf{n}$ is not zero, this gives

$$t = \frac{m - \mathbf{a} \cdot \mathbf{n}}{\mathbf{b} \cdot \mathbf{n}}$$

The position vector of P is the value of $\mathbf{r}$ at the point on the line with this parameter, namely

$$\mathbf{r} = \mathbf{a} + \mathbf{b}\,\frac{m - a \cdot n}{\mathbf{b} \cdot \mathbf{n}}$$

If $\mathbf{b} \cdot \mathbf{n} = 0$ then the line is parallel to the plane, in which case there is no intersection, or the line is in the plane.

Perpendicular Distance of a Plane from the Origin

Let the equation of the plane be $\mathbf{r} \cdot \mathbf{n} = m$. Then the required distance is the length of the orthogonal projection of $\mathbf{r}$ on $\mathbf{n}$. This is

$$\left| \frac{\mathbf{r} \cdot \mathbf{n}}{n} \right| = \left| \frac{m}{n} \right|$$

where the modulus is necessary in case m is negative. If, in particular, the equation of the plane is expressed as $\mathbf{r} \cdot \hat{\mathbf{n}} = d$, where $\hat{\mathbf{n}}$ is the unit vector perpendicular to the plane, then the distance of the plane from the origin is just $|d|$.

Parametric Equation of Plane

Let us construct the equation of the plane containing the three points A, B and C with position vectors $\mathbf{a}$, $\mathbf{b}$ and $\mathbf{c}$, respectively. We note that the plane is only defined if the three points are not collinear, that is, if they do not all lie on one line. Assuming that this is the case, we see that the lines AB and AC must lie in the plane (see Figure 5.12). Now consider the general point P in the plane having position vector $\mathbf{r}$. Draw a line from P parallel to AB, to intersect AC (possibly produced) in Q. Then the triangle law applied to

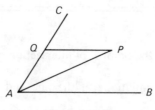

Fig. 5.12

AQP gives $\overrightarrow{AP} = \overrightarrow{AQ} + \overrightarrow{QP}$. But $\overrightarrow{AQ}$ is a multiple, s, say, of $\overrightarrow{AC} = \mathbf{c} - \mathbf{a}$ and $\overrightarrow{QP}$ is a multiple, t, say, of $\overrightarrow{AB} = \mathbf{b} - \mathbf{a}$. We thus have

$$\overrightarrow{AP} = s(\mathbf{c} - \mathbf{a}) + t(\mathbf{b} - \mathbf{a})$$

The position vector of P is now given by

$$\mathbf{r} = \mathbf{a} + \overrightarrow{AP}$$

$$= \mathbf{a} + s(\mathbf{c} - \mathbf{a}) + t(\mathbf{b} - \mathbf{a})$$

This form is sometimes more useful than the previous form, $\mathbf{r} \cdot \mathbf{n} = m$.

Example 5.8.4

Find the length of the perpendicular from the origin to the plane containing the points A, B and C whose position vectors are $(-1, 0, 0)$, $(1, 2, 3)$ and $(0, -1, 1)$, respectively.

The neatest way of solving this problem is to express the equation of the plane in the form $\mathbf{r} \cdot \hat{\mathbf{n}} = d$, where $\hat{\mathbf{n}}$ is a unit vector perpendicular to the plane. First, we need a vector perpendicular to the plane; such a vector is $\mathbf{n} = \overrightarrow{AB} \times \overrightarrow{AC}$. We have $\overrightarrow{AB} = \overrightarrow{OB} - \overrightarrow{OA} = (2, 2, 3)$ and $\overrightarrow{AC} = \overrightarrow{OC} - \overrightarrow{OA} = (1, -1, 1)$. Using the determinant form to evaluate the vector product, we have

$$\mathbf{n} = \begin{vmatrix} \mathbf{i} & \mathbf{j} & \mathbf{k} \\ 2 & 2 & 3 \\ 1 & -1 & 1 \end{vmatrix}$$

$$= (5, 1, -4)$$

giving

$$\hat{\mathbf{n}} = \frac{(5, 1, -4)}{\sqrt{42}}$$

The equation of the plane is $\mathbf{r} \cdot \hat{\mathbf{n}} = d$, but, since the point A lies in the plane, this equation must be satisfied with $\mathbf{r} = (-1, 0, 0)$, which gives

$$d = (-1, 0, 0) \cdot (5, 1, -4) \frac{1}{\sqrt{42}}$$

$$= -\frac{5}{\sqrt{42}}$$

and so the required distance is $5/\sqrt{42}$.

EXERCISES 5.8.1

1 Find the cartesian equations of the line in Example 5.8.1.
2 Find the parametric equations of the straight lines which

  (i) pass through the points with position vectors (5, 1, 2) and (7, 2, 5);
  (ii) pass through the points with position vectors $(-9, 6, -4)$ and $(-5, 4, -3)$.

  Find the point of intersection of the lines in (i) and (ii).
3 A straight line passes through the points $(-1, 3, 4)$ and $(8, -3, -11)$. Another line passes through the points $(3, 5, 2)$ and $(7, 1, -2)$.

  (i) Find the parametric equations of these two lines.
  (ii) These lines are intersected at points P, Q by a straight line perpendicular to both of them. Write down expressions for $\overrightarrow{OP}$ and $\overrightarrow{OQ}$, where O is the origin, and then use the property that $\overrightarrow{PQ}$ is perpendicular to both lines to find the values of the parameters corresponding to P and Q.
  (iii) Find the distance PQ and the parametric equation of the line through P and Q.
4 Find the equation of the plane which passes through the point $(2, -2, 3)$ and is parallel to the plane $\mathbf{r} \cdot (2, 1, -3) = 4$. Find also the perpendicular distance between the planes.
5 Find the angle between the planes $\mathbf{r} \cdot (1, 1, 2) = 4$ and $\mathbf{r} \cdot (2, 1, 1) = 5$. (The angle between the planes equals the angle between the normals to the planes.)
6 Find the parametric equation of the plane through the points A, B, C with position vectors $\mathbf{a} = (1, -1, 3)$, $\mathbf{b} = (1, 5, 3)$, $\mathbf{c} = (-3, -7, 5)$. Use this expression to find the parametric equation of its line of intersection with the plane through the points with position vectors $\mathbf{x} = (4, 5, 3)$, $\mathbf{y} = (8, 5, 1)$, $\mathbf{z} = (5, 8, 4)$.

5.9 VECTOR EQUATIONS OF CURVES IN SPACE

Example 5.9.1

Suppose that a ball is thrown with an initial velocity $\mathbf{v}_0$. Find the trajectory and velocity of the subsequent motion of the ball.

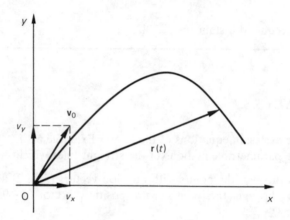

Fig. 5.13

Choose the coordinate system as in Figure 5.13, so that $\mathbf{v}_0$ is in the xy plane, with y vertically upwards and the origin at the launching point. If $\mathbf{v}_0 = (v_x, v_y)$, then the position vector of the ball at time t is given by

$$\mathbf{r}(t) = (x(t), y(t)) = (v_x t, v_y t - \tfrac{1}{2}gt^2)$$

Here $\mathbf{r}$ is an example of a vector function of t. We can write it in terms of the usual unit vectors,

$$\mathbf{r}(t) = v_x t \mathbf{i} + (v_y t - \tfrac{1}{2}gt^2)\mathbf{j}$$

The trajectory of the ball is a plane curve, a parabola in this case. In more general situations, such as flying a kite, the motion may not stay in a plane and we should need all three coordinates to describe the motion.

We give the following definition, which holds in a two- or three-dimensional context.

Definition

A *vector-valued function* $\mathbf{r}$ of t is a rule which associates with each t in a given interval a vector $\mathbf{r}(t)$.

In a three-dimensional setting, we can write this in component form

$$\mathbf{r}(t) = (x(t), y(t), z(t))$$
$$= x(t)\mathbf{i} + y(t)\mathbf{j} + z(t)\mathbf{k}$$

where x, y, z are functions of the variable t. This equation is called the *parametric equation* of the curve given by the locus of points $(x(t), y(t), z(t))$ as t takes values in the domain of $\mathbf{r}$. If one component is fixed, the curve becomes a *plane curve*, or a curve in two dimensions. For example, if $z(t) = 0$, then we have a curve in the xy plane.

In practice, the functions x, y, z will be well-behaved; in fact, they will usually belong to our family of standard functions. It is often convenient to think of $\mathbf{r}(t)$ as the position vector of a moving particle at time t, as in Example 5.9.1.

We have already constructed the equation of a straight line in Section 5.8. We now give examples of some more complex curves.

Examples 5.9.2

1. What curve is represented by the equation

$$\mathbf{r}(t) = (a \cos t, a \sin t)?$$

Since $x = a \cos t$ and $y = a \sin t$, we have $x^2 + y^2 = a^2 (\cos^2 t + \sin^2 t) = a^2$, so that the equation represents a circle of radius a, centred at the origin, as shown in Figure 5.14. We note that t is the angle that the vector $\mathbf{r}(t)$ makes with the positive x axis and that, as t increases, $\mathbf{r}(t)$ traces out the circle in an anticlockwise direction. The whole circle is traced out if t goes from 0 to 2π.

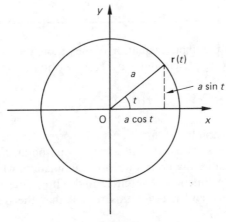

Fig. 5.14

2. What curve is represented by the equation

$$\mathbf{r}(t) = (a \cos t, \, a \sin t, \, bt), \, b > 0, \, t > 0?$$

As in the previous example, x and y satisfy $x^2 + y^2 = a^2$, but now z can also vary, so that x and y lie on a cylinder of radius a, whose axis is the z axis, as shown in Figure 5.15. As t increases from zero, $\mathbf{r}(t)$ moves round the cylinder and in the direction of increasing z. This curve is called a *circular helix*.

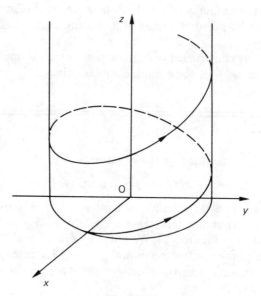

Fig. 5.15

We can find the parametric equation of the graph, $y = f(x)$, of the function f simply by replacing x by t and y by $f(t)$, to obtain

$$\mathbf{r}(t) = (t, \, f(t))$$

Note that the graphs of functions are special curves, since any line $x = $ constant meets such a curve in at most one point, for otherwise an x value would give more than one value for the function, which would violate our definition of a function. The circle and ellipse are, therefore, not the graphs of functions, and it is for this reason that the parametric form of equation is so useful.

EXERCISES 5.9.1

1 Define three vector functions

$$\mathbf{r}(t) = \mathbf{i} \cos t + \mathbf{j} \sin t, \ t \in [0, 2\pi]$$

$$\mathbf{r}_1(t) = \mathbf{i} \cos 2t + \mathbf{j} \sin 2t, \ t \in [0, \pi]$$

$$\mathbf{r}_2(t) = \mathbf{i} \cos(-t) + \mathbf{j} \sin(-t), \ t \in [0, 2\pi]$$

State whether the curve $\mathbf{r}$ is the same as the curve $\mathbf{r}_1$ or the curve $\mathbf{r}_2$. State also whether the function $\mathbf{r}$ is the same as the function $\mathbf{r}_1$ or the function $\mathbf{r}_2$.

2 Find the parametric equation of the circle of radius 3 and centre $(1, 3)$, directed counterclockwise.

3 Sketch and describe the curve in the xy plane given by

$$\mathbf{r}(t) = \mathbf{i}a \cos t + \mathbf{j}b \sin t, \ a, b > 0, \ t \in [0, 2\pi]$$

4 Sketch the following curves:

  (i) $\mathbf{r}(t) = \mathbf{i}a \cos t + \mathbf{j}b \sin t + \mathbf{k}t, \ a, b > 0, \ t \in \mathbb{R}$;

  (ii) $\mathbf{r}(t) = \mathbf{i}t \cos t + \mathbf{j}t \sin t, \ t \in [0, \infty)$;

  (iii) $\mathbf{r}(t) = \mathbf{i}t^3 + \mathbf{j}t, \ t \in \mathbb{R}$;

  (iv) $\mathbf{r}(t) = \mathbf{i} \cosh t + \mathbf{j} \sinh t, \ t \in \mathbb{R}$;

  (v) $\mathbf{r}(t) = \mathbf{i} \cos 2t \cos t + \mathbf{j} \cos 2t \sin t, \ t \in [0, 2\pi]$.

5.10 DIFFERENTIATION OF VECTOR FUNCTIONS

If we regard the vector

$$\mathbf{r}(t) = (x(t), y(t), z(t))$$

as the position vector of a moving particle at time t, then the components of its velocity are $x'(t)$, $y'(t)$ and $z'(t)$ and may be put together to form the *velocity vector*

$$\mathbf{v}(t) = (x'(t), y'(t), z'(t)).$$

Definition

Let the vector function $\mathbf{r}$ be given by

$$\mathbf{r}(t) = (x(t), y(t), z(t))$$

in some coordinate system. Then the *derivative*, $\mathbf{r}'$, is defined in the same coordinate system by

$$\frac{\mathrm{d}\mathbf{r}(t)}{\mathrm{d}t} = \mathbf{r}'(t) = (x'(t), y'(t), z'(t))$$

provided that x', y' and z' all exist.

It is clear, from the representation of $\mathbf{r}(t)$ as the position vector of a moving particle, that this derivative must be independent of the coordinate system. This is true for the derivatives of all vector functions, as will be apparent from the geometric result in the following theorem.

Theorem 5.10.1

Let $\mathbf{r}$ be a vector function in a plane given by $\mathbf{r}(t) = (x(t), y(t))$. If the vector $\mathbf{r}'(t_0)$ is not zero, then it is in the direction of the tangent at the point $\mathbf{r}(t_0)$ to the curve represented by the parametric equation $\mathbf{r}(t) = (x(t), y(t))$.

Proof   Assume first that $x'(t_0) \neq 0$. Then the slope of the tangent at $\mathbf{r}(t_0)$ is*

$$\left. \frac{dy}{dx} \right|_{t=t_0} = \left. \frac{dy/dt}{dx/dt} \right|_{t=t_0} = \frac{y'(t_0)}{x'(t_0)}$$

But the vector $\mathbf{r}'(t_0) = (x'(t_0), y'(t_0))$ has slope

$$\frac{y'(t_0)}{x'(t_0)}.$$

If $x'(t_0) = 0$, then the curve is parallel to the y axis and this is also the direction of $\mathbf{r}'(t_0) = (0, y'(t_0))$. $\square$

Example 5.10.1

We cannot differentiate $y = x^{1/3}$ at $x = 0$, but the y axis is a tangent. A parametric form of this equation is $\mathbf{r}(t) = (t^3, t)$ and so $(3t^2, 1)$ is a vector in the direction of the tangent to the curve at $\mathbf{r}(t)$, which takes the value $(0, 1)$ at $(0, 0)$. This shows the advantage of the parametric representation for finding tangents in a case like this.

For space curves we can use the derivative to define tangents.

Definition

Let $\mathbf{r}$ be a vector function in space. If $\mathbf{r}'(t_0) \neq 0$ then the *tangent* at $t = t_0$ to the curve whose parametric equation is $\mathbf{r}(t) = (x(t), y(t))$ is the line through $\mathbf{r}(t_0)$ in the direction of $\mathbf{r}'(t_0)$. Any vector in this direction is said to be a *tangent vector of the curve* at t_0.

Given two vector functions $\mathbf{f}$ and $\mathbf{g}$, we can use vector operations to obtain new vector functions. For example, $\mathbf{f} + \mathbf{g}$ and $c\mathbf{f}$, where c is a scalar, and $\mathbf{f} \times \mathbf{g}$ are vector functions; $\mathbf{f} \cdot \mathbf{g}$ is a scalar function. The rules for differentiating these combinations are given in the following theorem.

---

* $\left. \dfrac{dy}{dx} \right|_{t=t_0}$ is a useful notation meaning that $\dfrac{dy}{dx}$ is evaluated at $t = t_0$.

Theorem 5.10.2

Let $\mathbf{f}$ and $\mathbf{g}$ be two differentiable vector functions, h a scalar function of t and c a scalar constant. Then the following derivatives exist and are given by:

(1) $(c\mathbf{f})' = c\mathbf{f}'$
(2) $(h\mathbf{f})' = h'\mathbf{f} + h\mathbf{f}'$
(3) $(\mathbf{f} + \mathbf{g})' = \mathbf{f}' + \mathbf{g}'$
(4) $(\mathbf{f} \cdot \mathbf{g})' = \mathbf{f}' \cdot \mathbf{g} + \mathbf{f} \cdot \mathbf{g}'$
(5) $(\mathbf{f} \times \mathbf{g})' = \mathbf{f}' \times \mathbf{g} + \mathbf{f} \times \mathbf{g}'$

Proof The proofs of (1)–(3) are quite straightforward. For (4), let

$$\mathbf{f} = (x_1, y_1, z_1) \quad \text{and} \quad \mathbf{g} = (x_2, y_2, z_2)$$

Then

$$
\begin{aligned}
(\mathbf{f} \cdot \mathbf{g})' &= (x_1 x_2 + y_1 y_2 + z_1 z_2)' \\
&= (x_1' x_2 + y_1' y_2 + z_1' z_2) + (x_1 x_2' + y_1 y_2' + z_1 z_2') \\
&= \mathbf{f}' \cdot \mathbf{g} + \mathbf{f} \cdot \mathbf{g}'
\end{aligned}
$$

Part (5) goes similarly. $\qquad\square$

Theorem 5.10.3

Let $\mathbf{f}$ be a differentiable vector function and assume that $|f(t)|$ is constant. Then $\mathbf{f}(t) \cdot \mathbf{f}'(t) = 0$.

Proof Let $|f(t)| = c$, where c is a constant. Then differentiating $\mathbf{f}(t) \cdot \mathbf{f}(t) = c^2$ gives

$$\mathbf{f}'(t) \cdot \mathbf{f}(t) + \mathbf{f}(t) \cdot \mathbf{f}'(t) = 0$$

which gives the result. $\qquad\square$

What this says is that if $\mathbf{f}(t)$ is the position vector of a moving particle which is at a constant distance from the origin (that is, it moves on the surface of a sphere), then the velocity vector is orthogonal to the radius vector.

EXERCISES 5.10.1

1 Find the equation of the tangent to the curve

   (i) in Example 5.9.1.3 at $t = \pi/2$;
   (ii) in Exercise 5.9.1.4(ii) at $t = 0$ and $t = \pi/4$;
   (iii) in Exercise 5.9.1.4(iv) at $t = 0$ and $t = \ln 4$.

2 Let $\mathbf{f}(t) = t\mathbf{i} + t^2\mathbf{j} + t^3\mathbf{k}$ and $\mathbf{g}(t) = t^{-3}\mathbf{i} + t^{-2}\mathbf{j} + t^{-1}\mathbf{k}$. Calculate $(\mathbf{f} - \mathbf{g})'(t)$ and $(\mathbf{f} \cdot \mathbf{g})'(t)$.

3 Let $\mathbf{f}(t) = \mathbf{i} \cos t + \mathbf{j} \sin t$. Calculate $\mathbf{f}(t) \cdot \mathbf{f}'(t)$ and verify that it is zero.

5.11 ARCLENGTH

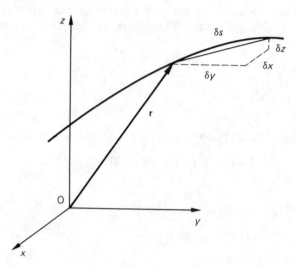

Fig. 5.16

Consider a moving particle, illustrated in Figure 5.16, with position vector $\mathbf{r} = (x, y, z)$, where x, y, z are functions of the time t. Suppose that at time $t + \delta t$ the particle has coordinates $(x + \delta x, y + \delta y, z + \delta z)$. The distance δs covered by the particle in time δt can be approximated by

$$\delta s = \sqrt{\delta x^2 + \delta y^2 + \delta z^2}$$

Dividing both sides of this equation by δt and taking the limit as $\delta t \to 0$ leads to

$$\frac{ds}{dt} = \sqrt{\left(\frac{dx}{dt}\right)^2 + \left(\frac{dy}{dt}\right)^2 + \left(\frac{dz}{dt}\right)^2}$$

which gives us the speed at time t of the particle along the curve. Since this is also given by $|\mathbf{r}'(t)|$, we have

$$\frac{ds}{dt} = |\mathbf{r}'(t)|$$

We note in passing that, since $d\mathbf{r}/dt$ is a vector in the direction of the tangent to the curve, then ds/dt is the length of this vector; thus, a unit vector in the direction of the tangent is

$$\frac{d\mathbf{r}}{dt} \bigg/ \frac{ds}{dt} = \frac{d\mathbf{r}}{ds}$$

To obtain the arclength s of the curve between $t = a$ and $t = b$, we simply integrate ds/dt, to obtain

$$s = \int_a^b \frac{ds}{dt}\, dt$$

$$= \int_a^b \sqrt{\left(\frac{dx}{dt}\right)^2 + \left(\frac{dy}{dt}\right)^2 + \left(\frac{dz}{dt}\right)^2}\, dt$$

$$= \int_a^b |\mathbf{r}'(t)|\, dt$$

Example 5.11.1

Find the arclength from $t = a$ to $t = b$ of the logarithmic spiral given by
$\mathbf{r}(t) = e^t \cos t\, \mathbf{i} + e^t \sin t\, \mathbf{j}$.
 Differentiating $\mathbf{r}$ gives

$$\mathbf{r}'(t) = e^t(\cos t - \sin t)\mathbf{i} + e^t(\sin t + \cos t)\mathbf{j}$$

so that

$$|\mathbf{r}'(t)| = \sqrt{e^{2t}(\cos t - \sin t)^2 + e^{2t}(\cos t + \sin t)^2}$$

$$= \sqrt{2}\, e^t$$

The required arclength is

$$\int_a^b \sqrt{2}\, e^t\, dt = \sqrt{2}(e^b - e^a)$$

EXERCISES 5.11.1

1 Calculate the length of the arc

(i) in Example 5.9.2.2 between $t = 0$ and $t = 2\pi$;
(ii) in Exercise 5.9.1.4(ii) between $t = 0$ and $t = 2\pi$.

2 Let f be a differentiable function. Show that the arclength of the graph
of f between $x = a$ and $x = b$ is given by the integral

$$\int_a^b \sqrt{1 + f'(x)^2}\, dx$$

3 Calculate the length of the parabola $y = x^2$ between $x = 0$ and $x = 2$.

3 Find the length of the circle $\mathbf{r}(t) = \mathbf{i} \cos t + \mathbf{j} \sin t$ from $t = 0$ to $t = a$ and interpret the result.

MISCELLANEOUS EXERCISES 5

1 Suppose that P has coordinates (x, y) with respect to the set of axes Oxy. Find its coordinates (x', y') with respect to the set of axes $Ox'y'$ formed by rotating Oxy anticlockwise through an angle θ. Show that the coordinates satisfy

$$\begin{bmatrix} x' \\ y' \end{bmatrix} = \begin{bmatrix} \cos\theta & \sin\theta \\ -\sin\theta & \cos\theta \end{bmatrix} \begin{bmatrix} x \\ y \end{bmatrix}$$

Show that the determinant of the *rotation matrix*,

$$\begin{bmatrix} \cos\theta & \sin\theta \\ -\sin\theta & \cos\theta \end{bmatrix}$$

has the value 1.

2 Suppose that the set of axes $Oxyz$ are rotated to form the set $Ox'y'z'$, where Ox' has direction cosines l_{11}, l_{12}, l_{13}, Oy' has direction cosines l_{21}, l_{22}, l_{23} and Oz' has direction cosines l_{31}, l_{32}, l_{33} with respect to $Oxyz$. Let A be the 3×3 matrix whose i, jth element is l_{ij} for $i, j = 1, 2, 3$. Show that $AA^T = I$ and deduce that $\det A = 1$.

3 Let P have coordinates (x, y, z) and (x', y', z') with respect to axes $Oxyz$ and $Ox'y'z'$ as in Exercise 2. Use the fact that x' is the length of the orthogonal projection of OP onto Ox', etc., to show that

$$\begin{bmatrix} x' \\ y' \\ z' \end{bmatrix} = \begin{bmatrix} l_{11} & l_{12} & l_{13} \\ l_{21} & l_{22} & l_{23} \\ l_{31} & l_{32} & l_{33} \end{bmatrix} \begin{bmatrix} x \\ y \\ z \end{bmatrix}$$

4 What rotation of axes is given by the matrix

$$\begin{bmatrix} \sin\theta\cos\phi & \sin\theta\sin\phi & \cos\theta \\ \cos\theta\cos\phi & \cos\theta\sin\phi & -\sin\theta \\ -\sin\phi & \cos\phi & 0 \end{bmatrix} ?$$

5 Let a disc of radius a roll along the x axis at a rate of 1 radian per second. Show that a point of the disc distance $b \leq a$ from the centre traces out the curve given by

$$\mathbf{r}(t) = (at - b \sin t, a - b \cos t)$$

Sketch the curve when $b < a$ (a *trochoid*) and when $b = a$ (a *cycloid*).

6 The equation of a straight line is

$$\mathbf{r}(t) = (1,\ 1,\ 1) + t(1,\ 2,\ 3)$$

Show that this may be written in *intrinsic* form as

$$\mathbf{r}(t) = (1,\ 1,\ 1) + \frac{s}{\sqrt{14}}(1,\ 2,\ 3)$$

where s is the arclength of the line from $(1, 1, 1)$.

7 Find the intrinsic form of equation of the circle

$$\mathbf{r}(t) = (a \cos t,\ a \sin t)$$

8 The unit tangent, that is, a unit vector in the direction of the tangent, of a curve traced out by $\mathbf{r}(s)$ is given by $\dfrac{d\mathbf{r}}{ds} = \mathbf{T}(s)$, say, where s is the arclength of the curve, and the curve is expressed in intrinsic form. Then the *curvature* κ and *radius of curvature* ρ are defined by

$$\kappa(s) = \frac{1}{\rho(s)} = \left| \frac{d\mathbf{T}}{ds} \right|$$

Since $\mathbf{T}$ is a unit vector, $\dfrac{d\mathbf{T}}{ds}$ is perpendicular to $\mathbf{T}$ (Theorem 5.10.3) and, hence, in the direction of the normal. The *principal unit normal* $\mathbf{N}(s)$ is defined by

$$\mathbf{N}(s) = \frac{\dfrac{d\mathbf{T}}{ds}}{\left| \dfrac{d\mathbf{T}}{ds} \right|} = \rho(s) \frac{d\mathbf{T}}{ds}$$

Verify that the radius of curvature of the circle in Exercise 7 is a and that its principal unit normal points to its centre.

9 For the curve $\mathbf{r}(x) = (x, y)$, where y is a function of x, show that

$$\mathbf{T}(x) = \frac{(1, y')}{\sqrt{1 + y'^2}}, \quad \kappa(x) = \frac{|y''|}{|(1 + y'^2)^{3/2}|} \quad \text{and} \quad \mathbf{N}(x) = \frac{(-y', 1)}{\sqrt{1 + y'^2}}$$

10 Show that the area of the triangle whose vertices have coordinates $(x_1, y_1), (x_2, y_2), (x_3, y_3)$ is

$$\pm \frac{1}{2} \begin{vmatrix} x_1 & y_1 & 1 \\ x_2 & y_2 & 1 \\ x_3 & y_3 & 1 \end{vmatrix}$$

5.12 ANSWERS TO EXERCISES

Exercises 5.2.1

1 $\sqrt{8^2 + 4^2 + 1^2} = 9$.

2 Let (x, y, z) be such a point, then $y^2 + z^2 = \frac{1}{2}$, $z^2 + x^2 = \frac{1}{2}$ and $x^2 + y^2 = \frac{1}{2}$. Subtracting the first two of these equations gives $y^2 = x^2$ and together with the third this gives $x^2 = \frac{1}{4}$ so $x = \pm \frac{1}{2}$; also y and z take the values $\pm \frac{1}{2}$. The complete solution consists of $(\pm \frac{1}{2}, \pm \frac{1}{2}, \pm \frac{1}{2})$ with all possible sign combinations.

3 If Q has coordinates (x, y, z) with respect to $O'xyz$, then it has coordinates $(x + 1, y + 1, z - 1)$ with respect to $Oxyz$; if now Q is the point O, then $(x + 1, y + 1, z - 1) = (0, 0, 0)$, so $(x, y, z) = (-1, -1, 1)$ gives the co-ordinates of O with respect to $O'xyz$.

Taking $Q = P$ with coordinates $(-1, 2, 0)$ with respect to $O'xyz$, then as above the coordinates with respect to $Oxyz$ are $(-1 + 1, 2 + 1, 0 - 1) = (0, 3, -1)$.

Exercises 5.3.1

1 $(-8, 5, 0)$, $(-5, -3, -7)$, $\sqrt{\frac{21}{2}}$.

$\overrightarrow{PB} = \frac{1}{2}\overrightarrow{AB}$, $\overrightarrow{BQ} = \frac{1}{2}\overrightarrow{BC}$, so $\overrightarrow{PQ} = 2\overrightarrow{PB} + \overrightarrow{BQ} = \frac{1}{2}(\overrightarrow{AB} + \overrightarrow{BC}) = \frac{1}{2}\overrightarrow{AC}$. Similarly, $\overrightarrow{SR} = \frac{1}{2}\overrightarrow{AC}$, so SR is parallel to PQ and equal in length, and so $PQRS$ is a parallelogram.

3 (i) $\overrightarrow{AX} = \lambda\overrightarrow{AC} = \lambda(\mathbf{a} + \mathbf{b})$ by the parallelogram law.

(ii) $\overrightarrow{AX} = \overrightarrow{AD} + \overrightarrow{DX} = \mathbf{b} + \mu\overrightarrow{DM} = \mathbf{b} + \mu(\frac{1}{2}\mathbf{a} - \mathbf{b})$.

(iii) Equating the two versions of $\overrightarrow{AX}$ gives $\lambda\mathbf{a} + \lambda\mathbf{b} = \mathbf{b} + \mu(\frac{1}{2}\mathbf{a} - \mathbf{b})$, or, rearranging, $(\lambda - \frac{1}{2}\mu)\mathbf{a} = (1 - \lambda - \mu)\mathbf{b}$. Assuming that $\mathbf{a}$ and $\mathbf{b}$ are not in the same direction, this must mean that $\lambda = \frac{1}{2}\mu$ and $\lambda + \mu = 1$, which gives $\lambda = \frac{1}{3}$ as required.

Exercises 5.4.1

1 If θ is the angle between L_1 and L_2, then

$$\cos \theta = \frac{\sqrt{3}}{2} \cdot 0 + \frac{1}{2} \cdot 0 - 1 \cdot 0 = 0,$$

so the lines are perpendicular. Let (l, m, n) be the direction cosines of L_3. Then

$$\frac{\sqrt{3}}{2} l + \frac{1}{2} m = 0 \quad \text{and} \quad -n = 0.$$

But $l^2 + m^2 + n^2 = 1$, so

$$l = \pm\frac{1}{2}, \quad m = \mp\frac{\sqrt{3}}{2}.$$

Thus L_3 has direction cosines

$$\frac{1}{2}, -\frac{\sqrt{3}}{2}, 0 \quad \text{or} \quad -\frac{1}{2}, \frac{\sqrt{3}}{2}, 0$$

2 Let the cube have sides of length 2; consider the diagonal joining the vertices $(-1, -1, -1)$ and $(1, 1, 1)$, which has direction cosines

$$\frac{2}{\sqrt{2^2 + 2^2 + 2^2}} = \frac{1}{\sqrt{3}}, \frac{1}{\sqrt{3}}, \frac{1}{\sqrt{3}}$$

and the diagonal joining the vertices $(-1, -1, 1)$ and $(1, 1, -1)$, which has direction cosines $\frac{1}{\sqrt{3}}, \frac{1}{\sqrt{3}}, -\frac{1}{\sqrt{3}}$. Then if θ is the angle between them, $\cos\theta = \frac{1}{3} + \frac{1}{3} - \frac{1}{3} = \frac{1}{3}$, so $\theta = \cos^{-1}\frac{1}{3}$.

Exercises 5.5.1

1 Let θ be the required angle. Then

$$\cos\theta = \frac{(0, -1, 1)\cdot(3, 4, 5)}{\sqrt{0^2 + (-1)^2 + 1^2}\sqrt{3^2 + 4^2 + 5^2}} = \frac{1}{\sqrt{2}\sqrt{50}} = \frac{1}{10},$$

so $\theta = \cos^{-1}\frac{1}{10}$.

2 If the two vectors are perpendicular then $(1, 8, 2)\cdot(2\lambda^2, -\lambda, 4) = 0$, that is $2\lambda^2 - 8\lambda + 8 = 0$ or $(\lambda - 2)^2 = 0$, so that $\lambda = 2$.

If $\lambda = 2$ then the second vector is $(8, -2, 4)$ and then $(1, 8, 2)\cdot(8, -2, 4) = 0$, so the vectors are perpendicular.

3 $\vec{CB} = \mathbf{b} - \mathbf{c}$, and since OA is perpendicular to CB, $\mathbf{a}\cdot(\mathbf{b} - \mathbf{c}) = 0$. Similarly, $\mathbf{b}\cdot(\mathbf{a} - \mathbf{c}) = 0$ because $OB \perp CA$. Now $\mathbf{a}\cdot(\mathbf{b} - \mathbf{c}) = \mathbf{b}\cdot(\mathbf{a} - \mathbf{c}) \Rightarrow \mathbf{a}\cdot\mathbf{b} - \mathbf{a}\cdot\mathbf{c} = \mathbf{b}\cdot\mathbf{a} - \mathbf{b}\cdot\mathbf{c} \Rightarrow \mathbf{a}\cdot\mathbf{c} = \mathbf{b}\cdot\mathbf{c} \Rightarrow \mathbf{c}\cdot(\mathbf{a} - \mathbf{b}) = 0$. This says that $OC \perp BA$ and hence that the altitudes of a triangle pass through one point.

Exercises 5.6.1

1 $\begin{vmatrix} \mathbf{i} & \mathbf{j} & \mathbf{k} \\ a & b & 1 \\ 2 & 1 & 5 \end{vmatrix} = (5b - 1, -5a + 2, a - 2b) = (1, 3, -1)$. This gives $a = -\frac{1}{5}$, $b = \frac{2}{5}$ from the first two components, and the third component is correctly given with these values.

2 For example, $\mathbf{a} = (1, 1, 1)$, $\mathbf{b} = (1, 0, 0)$, $\mathbf{c} = (1, 0, 0)$. Then $\mathbf{a} \times (\mathbf{b} \times \mathbf{c}) = 0$ but $(\mathbf{a} \times \mathbf{b}) \times \mathbf{c} = (0, -1, -1)$.

3 Let $u = (a, b, c)$, then $(-b - c, a + 2c, a - 2b) = (0, 1, 1)$, so $b + c = 0$, $a + 2c = 1$, $a - 2b = 1$. Letting $c = \lambda$ we find $b = -\lambda$, $a = 1 - 2\lambda$ and $(1 - 2\lambda, -\lambda, \lambda)$ is the required solution for any value of λ.

4 $\mathbf{a} \times \mathbf{b}$ is perpendicular to the plane of $\mathbf{a}$ and $\mathbf{b}$, while $\mathbf{a} - \mathbf{b}$ is in the plane of $\mathbf{a}$ and $\mathbf{b}$. Thus both must be zero, and so $\mathbf{a} = \mathbf{b}$.

Exercises 5.7.1

1 It has been shown that $\cdot$ and $\times$ are interchangeable in a triple product and $\mathbf{r} \cdot \mathbf{s} = \mathbf{s} \cdot \mathbf{r}$. Thus $\mathbf{a} \cdot \mathbf{b} \times \mathbf{c} = \mathbf{a} \times \mathbf{b} \cdot \mathbf{c} = \mathbf{c} \cdot \mathbf{a} \times \mathbf{b} = \mathbf{c} \times \mathbf{a} \cdot \mathbf{b} = \mathbf{c} \cdot \mathbf{a} \times \mathbf{b}$.

2 $\mathbf{a} \times \mathbf{b}$ is perpendicular to $\mathbf{a}$ and $\mathbf{b}$, so its scalar product with $\mathbf{a}$ or $\mathbf{b}$ is zero. Also $\mathbf{a} \times \mathbf{a} \cdot \mathbf{b}$ is zero since $\mathbf{a} \times \mathbf{a}$ is zero.

3 If $\mathbf{a}$, $\mathbf{b}$, $\mathbf{c}$ are coplanar then $\mathbf{b} \times \mathbf{c}$ is perpendicular to all vectors in the plane of $\mathbf{b}$ and $\mathbf{c}$, which also contains $\mathbf{a}$. Hence $\mathbf{a} \cdot \mathbf{b} \times \mathbf{c} = 0$.

   If $\mathbf{a} \cdot \mathbf{b} \times \mathbf{c} = 0$ and none of $\mathbf{a}$, $\mathbf{b}$, $\mathbf{c}$ is zero, then either $\mathbf{a}$ is perpendicular to $\mathbf{b} \times \mathbf{c}$, in which case $\mathbf{a}$, $\mathbf{b}$, $\mathbf{c}$ are coplanar, or $\mathbf{b}$ is parallel to $\mathbf{c}$ and so only one plane is defined by the three vectors.

4 (i) Let $\mathbf{c} \times \mathbf{a} = \mathbf{d}$, then

$$\mathbf{a} \times (\mathbf{b} \times (\mathbf{c} \times \mathbf{a})) = \mathbf{a} \times (\mathbf{b} \times \mathbf{d})$$
$$= (\mathbf{a} \cdot \mathbf{d})\mathbf{b} - (\mathbf{a} \cdot \mathbf{b})\mathbf{d}$$
$$= (\mathbf{a} \cdot \mathbf{c} \times \mathbf{a})\mathbf{b} - (\mathbf{a} \cdot \mathbf{b})(\mathbf{c} \times \mathbf{a})$$
$$= (\mathbf{a} \cdot \mathbf{b})(\mathbf{a} \times \mathbf{c}).$$

(ii) With $\mathbf{d}$ as in (i),

$$(\mathbf{b} \times \mathbf{c}) \times (\mathbf{c} \times \mathbf{a}) = (\mathbf{b} \times \mathbf{c}) \times \mathbf{d}$$
$$= (\mathbf{b} \cdot \mathbf{d})\mathbf{c} - (\mathbf{c} \cdot \mathbf{d})\mathbf{b}$$
$$= (\mathbf{b} \cdot \mathbf{c} \times \mathbf{a})\mathbf{c} - (\mathbf{c} \cdot \mathbf{c} \times \mathbf{a})\mathbf{b}$$
$$= (\mathbf{a} \cdot \mathbf{b} \times \mathbf{c})\mathbf{c}.$$

5 Taking the vector product of the given equation with $\mathbf{a}$, we find $\lambda \mathbf{a} \times \mathbf{x} + \mathbf{a} \times (\mathbf{x} \times \mathbf{a}) = \mathbf{a} \times \mathbf{b}$. Also from the given equation, we have $\mathbf{x} \times \mathbf{a} = \mathbf{b} - \lambda\mathbf{x}$, and taking its scalar product with $\mathbf{a}$, $\lambda \mathbf{a} \cdot \mathbf{x} = \mathbf{a} \cdot \mathbf{b}$. Combining all these results, we find $(\lambda^2 + a^2)\mathbf{x} = \lambda\mathbf{b} + (\mathbf{a} \cdot \mathbf{b}/\lambda)\mathbf{a} + \mathbf{a} \times \mathbf{b}$.

Exercises 5.8.1

1 $\mathbf{r} = (x, y, z) = (1, 2, 3) + s(1, 1, 1)$, so $x = s + 1$, $y = s + 2$, $z = s + 3$, and eliminating the parameter s, $y = x + 1$, $z = y + 1$.

2 (i) $\mathbf{r} = (5, 1, 2) + u(2, 1, 3)$ and (ii) $\mathbf{r} = (-9, 6, -4) + v(4, -2, 1)$. At the point of intersection, these are equal, so $5 + 2u = -9 + 4v$, $1 + u = 6 - 2v$, $2 + 3u = -4 + v$, which give the values $u = -1$, $v = 3$, so the point of intersection is $(3, 0, -1)$.

3 (i) $\mathbf{r} = (-1, 3, 4) + u'(9, -6, -15) = (-1, 3, 4) + u(3, -2, -5)$,
$\mathbf{r} = (3, 5, 2) + v'(4, -4, -4) = (3, 5, 2) + v(1, -1, 1)$.

(ii) Let P, Q have parameters u, v, so that $\overrightarrow{OP} = (-1, 3, 4) + u(3, -2, -5)$
and $\overrightarrow{OQ} = (3, 5, 2) + v(1, -1, -1)$. Then $\overrightarrow{PQ} = \overrightarrow{OQ} - \overrightarrow{OP} = (3, 5, 2)$
$+ v(1, -1, -1) - (-1, 3, 4) - u(3, -2, -5) = (4, 2, -2) +$
$v(1, -1, -1) - u(3, -2, -5)$. But PQ is perpendicular to each of
the given lines, so

$(3, -2, -5)\overrightarrow{PQ} = 3(4 + v - 3u) - 2(2 - v + 2u) - 5(-2 - v + 5u) = 0$,

giving $38u - 10v = 18$ and $(1, -1, -1) \cdot \overrightarrow{PQ} = 0$, giving $10u - 3v = 4$.
The solution of these equations is $u = 1$, $v = 2$.

(iii) $\overrightarrow{PQ} = (3, 2, 1)$, so $PQ = \sqrt{14}$. The equation of PQ is

$$\mathbf{r} = (2, 1, -1) + \lambda(3, 2, 1).$$

4 The equation of a plane parallel to the given plane is $\mathbf{r} \cdot (2, 1, -3) = m$;
this contains the point $(2, -2, 3)$ if $m = (2, -2, 3) \cdot (2, 1, -3) = -7$. The

distances of the two planes from O are $\dfrac{4}{\sqrt{2^2 + 1^2 + 3^2}} = \dfrac{4}{\sqrt{14}}$ and $\dfrac{-7}{\sqrt{14}}$

so the distance between them is $\dfrac{11}{\sqrt{14}}$.

5 The required angle is given by $\cos\theta = \dfrac{(1, 1, 2) \cdot (2, 1, 1)}{\sqrt{(1^2 + 1^2 + 2^2)(2^2 + 1^2 + 1^2)}} = \dfrac{5}{6}$.

6 The plane containing A, B, C has parametric equation $\mathbf{r} = (1, -1, 3) +$
$s(0, 6, 0) + t(-4, -6, 2)$. The plane through $\mathbf{x}, \mathbf{y}, \mathbf{z}$ has parametric equation
$\mathbf{r} = (4, 5, 3) + \alpha(4, 0, -2) + \beta(1, 3, 1)$, and the intersection occurs where
$(1, -1, 3) + s(0, 6, 0) + t(-4, -6, 2) = (4, 5, 3) + \alpha(4, 0, -2) + \beta(1, 3, 1)$.

Solving these equations, we find $\beta = -1$, $\alpha = -\dfrac{1 + 2t}{2}$, $s = \dfrac{1 + 2t}{2}$, so the

intersection is given by

$$\mathbf{r} = (1, -1, 3) + \frac{1 + 2t}{2} (0, 6, 0) + t(-4, -6, 2) = (1, 2, 3) + t(-4, 0, 2).$$

Exercises 5.9.1

1 The curves are the same, although the last is traced out in the reverse
direction. The functions are not the same.

2 $\mathbf{r}(t) = (1, 3) + (3\cos t, 3\sin t) = \mathbf{i}(1 + 3\cos t) + \mathbf{j}(3 + 3\sin t), t \in [0, 2\pi]$.

3 $x = a\cos t, y = b\sin t$, so $\dfrac{x^2}{a^2} + \dfrac{y^2}{b^2} = 1$ and the required curve is an ellipse.

4 (i) This is an elliptic helix, and looks similar to the circular helix in Figure
5.15, but the cylinder on which it lies is elliptic instead of circular.

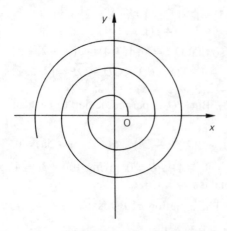

(ii) Spiral of Archimedes

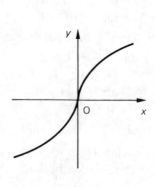

(iii) $y = x^{\frac{1}{3}}$

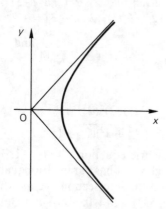

(iv) Rectangular hyperbola
$(x^2 - y^2 = 1)$

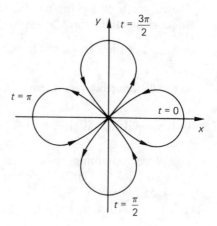

(v) 4-leaf clover

Exercises 5.10.1

1   (i) $\mathbf{r} = \left(0,\, a,\, \dfrac{\pi b}{2} \right) + \lambda(-a,\, 0,\, b).$

(ii) $\mathbf{r} = (0, 0) + \lambda(0, 1)$ at $t = 0$ and $\mathbf{r} = \dfrac{\pi}{4\sqrt{2}}(1, 1) + \lambda\left(\dfrac{4 - \pi}{4\sqrt{2}},\, \dfrac{4 + \pi}{4\sqrt{2}} \right)$ at

$t = \dfrac{\pi}{4}.$

(iii) $\mathbf{r} = (1, 0) + \lambda(0, 1)$ at $t = 0$ and $\mathbf{r} = \dfrac{1}{8}(17, 15) + \dfrac{\lambda}{8}(15, 17)$ at $t = \ln 4.$

2 $(\mathbf{f} - \mathbf{g})' = (1 + 3t^{-4}, 2t + 2t^{-3}, 3t^2 + t^{-2})$,
$(\mathbf{f} \cdot \mathbf{g})' = 2t(1 - t^{-4})$.
3 $\mathbf{f}(t) \cdot \mathbf{f}'(t) = (\cos t, \sin t) \cdot (-\sin t, \cos t) = 0$.

Exercises 5.11.1

1 (i) $2\pi\sqrt{1 + a^2}$,

(ii) $\int_0^{2\pi} \sqrt{1 + t^2}\, dt = [\frac{1}{2}\sinh^{-1} t + \frac{1}{2}t\sqrt{1 + t^2}]_0^{2\pi}$
$\qquad\qquad = \frac{1}{2}\{\ln(2\pi + \sqrt{1 + 4\pi^2}) + 2\pi\sqrt{1 + 4\pi^2}\}$.

2 In parametric form, $y = f(x)$ is $x = t$, $y = f(t)$, so

$$s = \int_a^b \sqrt{1 + f'(t)^2}\, dt = \int_a^b \sqrt{1 + f'(x)^2}\, dx.$$

3 $\int_0^2 \sqrt{1 + 4x^2}\, dx = \sqrt{17} + \frac{1}{4}\sinh^{-1} 4$, using the result of Question 1(ii) with $t = 2x$.

4 $s = a$, which is the arclength of a sector of the unit circle corresponding to an angle a.

Miscellaneous Exercises 5

1 $x' = OA' = OA \cos\theta + AP \sin\theta = x \cos\theta + y \sin\theta$,

$y' = A'P = AP \cos\theta - OA \sin\theta = y \cos\theta - x \sin\theta$,

and these satisfy the given matrix equation. The value of the determinant is $\cos^2\theta + \sin^2\theta = 1$.

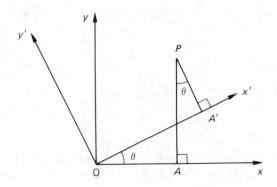

2

$$AA^T = \begin{bmatrix} l_{11} & l_{12} & l_{13} \\ l_{21} & l_{22} & l_{23} \\ l_{31} & l_{32} & l_{33} \end{bmatrix} \begin{bmatrix} l_{11} & l_{21} & l_{31} \\ l_{12} & l_{22} & l_{32} \\ l_{13} & l_{23} & l_{33} \end{bmatrix}$$

$$= \begin{bmatrix} \sum_{i=1}^3 l_{1i}^2 & \sum_{i=1}^3 l_{1i}l_{2i} & \sum_{i=1}^2 l_{1i}l_{3i} \\ \sum_{i=1}^3 l_{2i}l_{1i} & \sum_{i=1}^3 l_{2i}^2 & \sum_{i=1}^3 l_{2i}l_{3i} \\ \sum_{i=1}^3 l_{3i}l_{1i} & \sum_{i=1}^3 l_{3i}l_{2i} & \sum_{i=1}^3 l_{3i}^2 \end{bmatrix}$$

$$= I.$$

$1 = \det I = \det AA^T = \det A \cdot \det A^T = (\det A)^2 \Rightarrow \det A = \pm 1$. For no rotation, $A = I$, so $\det A = 1$; any small rotation cannot suddenly change the value of the determinant from 1 to -1, so its value must be $+1$. (-1 can only be obtained with a change to left-handed axes.)

3 The length of the projection of OP onto Ox′ is $\overrightarrow{OP} \cdot$ (unit vector along Ox′) $= (x, y, z) \cdot (l_{11}, l_{12}, l_{13}) = l_{11}x + l_{12}y + l_{13}z$. The projections of OP onto the other two axes give similar results.

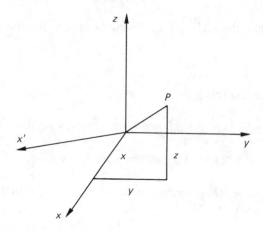

4 z' is in the xy plane at an angle ϕ to Oy. Ox' is at an angle θ to Oz.

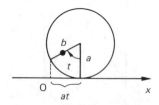

5 At t seconds; $\mathbf{r}(t) = (at - b\sin t, \ a - b\cos t)$. To check the shape of the curves, we find their slopes by differentiating. Thus

$\dot{\mathbf{r}}(t) = (a - b\cos t, b\sin t)$, so $\dfrac{dy}{dx} = \dfrac{b\sin t}{a - b\cos t}$. At $t = 0, 2\pi, 4\pi, \ldots, \dfrac{dy}{dx} = 0$,

if $a \neq b$. When $a = b$, however, $\dfrac{dy}{dx} = \dfrac{\sin t}{1 - \cos t} = \dfrac{\sin \frac{1}{2}t \cos \frac{1}{2}t}{\sin^2 \frac{1}{2}t} = \cot \frac{1}{2}t \to \infty$

when $t \to 0, 2\pi, 4\pi, \ldots$. The curves look like:

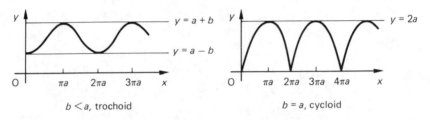

$b < a$, trochoid $\qquad\qquad\qquad\qquad\qquad$ $b = a$, cycloid

6 $s = t\sqrt{1^2 + 2^2 + 3^2} = t\sqrt{14}$ using Pythagoras's Theorem, so

$$\mathbf{r}(s) = (1,\ 1,\ 1) + \frac{s}{\sqrt{14}}\ (1,\ 2,\ 3).$$

7 The arclength of the circle is $s = at$ so $\mathbf{r}(s) = \left(a\cos\left(\dfrac{s}{a}\right),\ a\sin\left(\dfrac{s}{a}\right)\right)$.

8 For the circle in Question 7,

$$\mathbf{T}(s) = \frac{d\mathbf{r}}{ds} = \left(-\sin\left(\frac{s}{a}\right),\ \cos\left(\frac{s}{a}\right)\right),$$

so

$$\kappa(x) = \left|\frac{d\mathbf{T}}{ds}\right| = \left|\left(-\frac{1}{a}\cos\left(\frac{s}{a}\right),\ -\frac{1}{a}\sin\left(\frac{s}{a}\right)\right)\right|$$

$$= \frac{1}{a}, \text{ so } \rho(s) = a.$$

$$\mathbf{N}(s) = \left(-\cos\left(\frac{s}{a}\right),\ -\sin\left(\frac{s}{a}\right)\right),$$

which is a unit vector in the direction of the unit radius vector

$\left(\cos\left(\dfrac{s}{a}\right),\ \sin\left(\dfrac{s}{a}\right)\right)$.

9 $\mathbf{T}(s) = \dfrac{d\mathbf{r}}{ds} = \dfrac{d\mathbf{r}}{dx} \Big/ \dfrac{ds}{dx} = \dfrac{(1, y')}{\sqrt{1 + y'^2}}$.

$\dfrac{d\mathbf{T}}{ds} = \dfrac{d\mathbf{T}}{dx} \Big/ \dfrac{ds}{dx} = \left(\dfrac{-y'y''}{(1 + y'^2)^{3/2}},\ \dfrac{y''}{(1 + y'^2)^{3/2}}\right) \Big/ \sqrt{1 + y'^2}$, so $\kappa(s) = \dfrac{|y''|}{(1 + y'^2)^{3/2}}$.

173

Scaling the expression obtained above for $\dfrac{d\mathbf{T}}{ds}$ to be a unit vector, we

find $\mathbf{N}(s) = \dfrac{(-y', 1)}{\sqrt{1 + y'^2}}$.

10 Area of triangle $= \frac{1}{2}|(x_2 - x_1, y_2 - y_1) \times (x_3 - x_1, y_3 - y_1)|$
$= \pm \{(x_2 - x_1)(y_3 - y_1) - (y_2 - y_1)(x_3 - x_1)\}.$

Also

$$\pm \frac{1}{2} \begin{vmatrix} x_1 & y_1 & 1 \\ x_2 & y_2 & 1 \\ x_3 & y_3 & 1 \end{vmatrix} = \pm \frac{1}{2} \begin{vmatrix} x_1 & y_1 & 1 \\ x_2 - x_1 & y_2 - y_1 & 0 \\ x_3 - x_1 & y_3 - y_1 & 0 \end{vmatrix},$$

which by expanding about the third column gives the same expression as above.

6 FUNCTIONS OF TWO VARIABLES

6.1 INTRODUCTION

We have studied functions of one variable and vector functions of one variable. In each case, the value of the variable defines the value of the function. We now turn to functions of two variables, where the values of *two* variables must be given before we can evaluate the function. As in the case of a function of one variable, we define a function f of two variables, x and y, say, by its value at all points of the domain. Now, however, ranges of values for both x and y must be specified. For example, the domain of f might consist of values of x and y satisfying $a \leq x \leq b$ and $c \leq y \leq d$. Instead of referring to x and y separately, we shall often refer to them as a point (x, y); then we can talk about f being evaluated at the point (x, y), while the domain of f is the rectangle in the xy plane defined by the above inequalities.

Definitions

A *function f of two variables* x and y is a rule which associates with each point (x, y) a unique number $f(x, y)$.

The *domain* of f is the set of all points (x, y) for which $f(x, y)$ is defined.

The *range* of f is the set of all values of $f(x, y)$ as (x, y) varies through the domain of f.

Example 6.1.1

f is the function defined by $f(x, y) = xy + \ln(x + y)$ on the domain $x > 0, y > 0$. The range of f is $(-\infty, \infty)$.

We note that this function could not have been defined on a domain which included values of x and y for which $x + y \leq 0$, since then we could not have evaluated the logarithm. We could have given as an alternative domain that part of the xy plane for which $x + y > 0$.

6.2 THE STANDARD FAMILY OF FUNCTIONS

Just as in the one-variable case, we have a number of simple functions, such as constants, linear functions of x and y, together with exponentials, logarithms, and trigonometric and hyperbolic functions. These are combined to build up members of the *standard family of functions of two variables*, using a finite number of additions, multiplications and divisions and compositions. For example, the polynomial function given by

$$f(x, y) = 3x^3 y^2 + 2xy^2 - xy + x - 5$$

is made up from the functions $f(x, y) = x$, $f(x, y) = y$ and $f(x, y) = $ constant by multiplication, addition and subtraction. Including division, we can construct rational functions, such as

$$f(x, y) = \frac{3x^2 y - xy + 7}{x^3 + y^2}$$

In specifying the domain of this function, we should have to exclude all points at which the denominator vanishes.

Let the function h be defined by $h(x, y) = x^2 + y^3$ and the function g (of one variable only, note) by $g(t) = \sin t$. Now define f by

$$f(x, y) = g(h(x, y))$$

$$= \sin(h(x, y))$$

$$= \sin(x^2 + y^3)$$

Then f is a member of the standard family of functions of two variables. Its domain could be the whole of the xy plane, when its range would be $[-1, 1]$.

EXERCISE 6.2.1

Give the largest possible domain and the corresponding range for the functions given by

(i) $\dfrac{1}{x^2 + y^2}$;   (ii) $\dfrac{1}{x^2 + y^2 - 1}$;   (iii) $\ln(x - y)$.

6.3 GRAPHICAL REPRESENTATION

The graph of a function f of one variable consists of points in the xy plane with coordinates $(x, f(x))$, where x takes all possible values in the domain of f. The corresponding notion for a function f of two variables is the set of points in space with coordinates $(x, y, f(x, y))$, where (x, y)

ranges over all values in the domain of f. This set of points gives a *surface* in space. For example, as we saw in Section 5.2, the surface represented by $z = f(x, y) = r$ (a constant) is a plane parallel to the xy plane and distance r above it.

Example 6.3.1

Describe the surface represented by the equation

$$z = f(x, y) = \sqrt{x^2 + y^2}$$

for all values of (x, y).

The intersection of the surface with the plane $z = r$, where r is a constant, is $x^2 + y^2 = r^2$, which represents a circle of radius r. The radius increases linearly with z, so the surface is a cone, as shown in Figure 6.1.

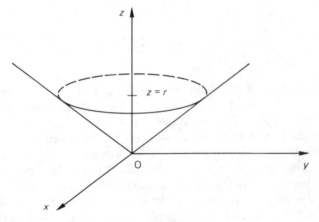

Fig. 6.1

In this example we were able to see what the surface looked like by considering the curves of intersection with planes $z = $ constant. These are called *level curves* of f. They are exactly what the mapmaker calls contours. If a number of level curves are drawn in the xy plane for equally spaced values of r, we can obtain insight into the shape of the surface. In particular, we get an indication of how steep the surface is at a certain point from the closeness of the level curves. Examples of level curves are shown in Figure 6.2.

Sometimes surfaces have certain symmetries; spotting these often eases the task of constructing them. For example, if the function f satisfies $f(-x, y) = f(x, y)$ for all values of x, then the surface $z = f(x, y)$ is symmetric about the yz plane.

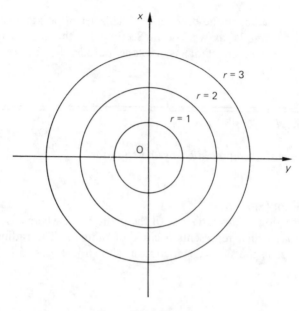

Fig. 6.2

Another useful way of visualising surfaces is to find their intersections with planes $x = $ constant or $y = $ constant; this will turn out to be very useful when we seek to differentiate functions of two variables.

We can go even further by considering the intersection of f with a surface defined in space by $y = g(x)$. It is perhaps not immediately clear why this equation does represent a surface. Since z does not appear, it must be understood that $y = g(x)$ is satisfied for *every* value of z; thus, the curve $y = g(x)$ is repeated in every plane $z = $ constant. The surface so defined is called a *cylinder*. For example, the equation $x^2 + y^2 = r^2$ defines a circle of radius r in every plane $z = $ constant, and so represents a circular cylinder of radius r, whose axis is the z axis. It is often convenient to give the equation of a cylinder in parametric form, $\mathbf{r}(t) = (x(t), y(t))$. The intersection of this cylinder with the surface given by $z = f(x, y)$ is then the curve in space whose equation in parametric form is

$$\mathbf{r}(t) = (x(t), y(t), f(x(t), y(t)))$$

Mostly we shall examine the intersection of surfaces $z = f(x, y)$ by cylinders of the form $y = ax + b$, that is, by planes; in this case the intersection is called the *cross-section* of f in the plane $y = ax + b$. Often it is sufficient for the visualisation of a surface just to consider cross-sections in planes $y = b$ and $x = a$. The intersection of $z = f(x, y)$ with $y = b$ is the curve $z = f(x, b)$, which is simply the graph of a one-variable function in the plane $y = b$.

Example 6.3.2

Let $z = f(x, y) = x^2 - y^2$. The cross-section of f in the plane $y = b$ is given by $z = x^2 - b^2$, which is a parabola in the plane $y = b$, illustrated in Figure 6.3.

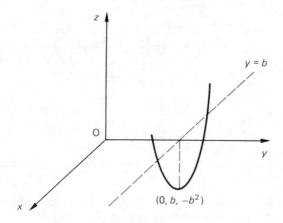

Fig. 6.3

As we take different values for b, so the parabola changes; to see how it changes, we look at the cross-section of f in the plane $x = a$, which is given by $z = a^2 - y^2$. This is also a parabola, as shown in Figure 6.4, but it is 'upside down' compared with the previous one. We can attempt to visualise the surface of f by drawing cross-sections for several values of a and b on the same picture. The result is shown in Figure 6.5. The part shown can be likened to a saddle.

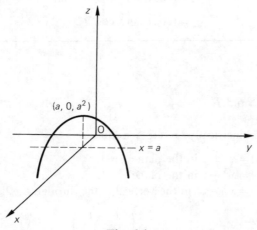

Fig. 6.4

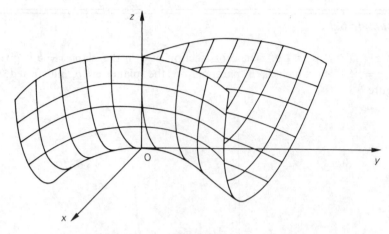

Fig. 6.5

It is no more difficult to find the curve of intersection of a surface with a non-planar cylinder, as the following example shows.

Example 6.3.3

Find the parametric equation of the curve of intersection of the surface $z = x^2 - y^2$ and the circular cylinder whose parametric equation is $\mathbf{r}(t) = (\cos t, \sin t)$.

This is simply

$$\mathbf{r}(t) = (\cos t, \sin t, \cos^2 t - \sin^2 t)$$

$$= (\cos t, \sin t, \cos 2t).$$

EXERCISES 6.3.1

1 Describe geometrically the cross-section of:

   (i) $f(x, y) = x^3 + y$ in the plane $x = 1$,

   (ii) $f(x, y) = x^3 + y$ in the plane $y = 1$,

   (iii) $f(x, y) = x^2 - y^2$ in the vertical plane through the line

$$\left(\frac{t}{\sqrt{2}}, \frac{t}{\sqrt{2}} \right),$$

(iv) $f(x, y) = x^2 - y^2$ in the vertical plane through the line

$$\left(\frac{t}{2}, \frac{t\sqrt{3}}{2} \right).$$

2 Sketch the surface of $z = x^2 + y^2$. Describe the level curves of this surface.

3 Find a tangent vector at $t = \pi/4$ of the curve in Example 6.3.3.

4 Show that the curve $(t \cos t, t \sin t, t^2)$, $t \geq 0$, lies on the surface $z = x^2 + y^2$. Describe this curve.

5 Describe geometrically the curve of intersection of the surface $z = x^2 - y^2$ and the cylinder $\mathbf{r}(t) = (\cosh t, \sinh t)$.

6.4 FUNCTIONS OF THREE OR MORE VARIABLES

We shall sometimes wish to work with functions of three variables, and we can include the case of more than three variables, since no new principle would be involved in this. Standard families of functions of three or more functions are built up in the same way as those for two variables.

Just as we need functions of two variables to define conveniently certain curves, such as a circle whose equation is $f(x, y) = x^2 + y^2 = r^2$, so we need functions of three variables to define some three-dimensional surfaces. Examples include $f(x, y, z) = x^2 + y^2 + z^2 = r^2$, which is the equation of a sphere of radius r, centred at the origin.

We note that a surface given in the normal way by $z = f(x, y)$ can be expressed in the form $g(x, y, z) = 0$, where $g(x, y, z) = f(x, y) - z$.

6.5 PARTIAL DERIVATIVES

Suppose that the equation of a mountain is given by $z = f(x, y)$ and that we want to choose a route to the top from our present position, which is the point $(x', y', f(x', y'))$. To make such a choice, we should want information about the slope of the mountain in various directions. The level curves give some information about this, but it may not be precise enough. We can obtain further information by computing the slopes of cross-sectional curves of f in planes $x = x'$ and $y = y'$, which are given by $z = f(x', y)$ and $z = f(x, y')$, respectively. The slope of the latter at the point (x', y') is obtained by differentiating $f(x, y')$ with respect to x and putting $x = x'$. We write the result as $\partial f(x', y')/\partial x$ or $f_x(x', y')$. If we now let (x', y') range through all points in the xy plane, we obtain a new function of two variables, f_x, called the *partial derivative* of f with respect to x. For example, if $f(x, y) = x^2 - y^2$, then $f(x, y') = x^2 - y'^2$. Differentiating with respect to x and putting $x = x'$ gives $f_x(x', y') = 2x'$. The partial derivative of f with respect to x is thus given by $f_x(x, y) = 2x$. We could have obtained this result more directly by just differentiating f with respect to x while treating y as a constant. When we are given $z = f(x, y)$, we shall often write $\partial z/\partial x$ in place of f_x. Note that ∂ is a special symbol and not d or a greek δ. The partial derivative of f with respect to y, $\partial f/\partial y$ or f_y, is found in a similar way, that is, by differentiating with respect to y, while treating x as a constant.

Example 6.5.1

Find f_x and f_y in the following cases:

(i) $f(x, y) = x^3 - y^4$

Keeping y constant and differentiating with respect to x, we obtain $f_x(x, y) = 3x^2$. Similarly, keeping x constant and differentiating with respect to y gives $f_y(x, y) = -4y^3$.

(ii) $f(x, y) = \sin xy$

$$f_x(x, y) = y \cos xy \text{ and } f_y(x, y) = x \cos xy.$$

(iii) $f(x, y) = x^2/y^3$

$$f_x(x, y) = 2x/y^3 \text{ and } f_y(x, y) = -(3x^2/y^4).$$

We can obtain partial derivatives of functions of three variables x, y and z by keeping two of the variables constant while differentiating with respect to the third.

Example 6.5.2

Let $f(x, y, z) = x^2 yz - x + yz^2$.

$$f_x(x, y, z) = 2xyz - 1, f_y(x, y, z) = x^2z + z^2 \text{ and } f_z(x, y, z) = x^2 y + 2yz.$$

We end this section by giving formal definitions of the partial derivatives of a function f of two variables. They are modelled on those for a function of one variable.

Definitions

Suppose that f is defined on a domain which contains (x, y). Then the partial derivatives f_x and f_y are given by

$$f_x(x, y) = \lim_{\delta x \to 0} \frac{f(x + \delta x, y) - f(x, y)}{\delta x}$$

and

$$f_y(x, y) = \lim_{\delta y \to 0} \frac{f(x, y + \delta y) - f(x, y)}{\delta y}$$

if the limits exist.

These definitions clearly confirm the method of obtaining partial derivatives by differentiating with respect to one variable while keeping the other constant.

EXERCISES 6.5.1

1 Find the partial derivatives with respect to x and y of the functions given by:

(i) $f(x, y) = x^2 + y^2$; (ii) $f(x, y) = \dfrac{1}{x + y}$;

(iii) $f(x, y) = \ln(x^2 + y^2 + 1)$; (iv) $f(x, y) = \tan^{-1} \dfrac{y}{x}$;

(v) $f(x, y) = xe^{xy^2}$; (vi) $f(x, y) = x^y \ (x > 0)$.

2 Find the partial derivatives with respect to x, y and z of the functions given by

(i) $f(x, y, z) = z \sin(yz^3 + x)$; (ii) $f(x, y, z) = \ln(xy^2z^3)$.

6.6 CHAIN RULES

We now consider what happens when f is a function of x and y, and x and y are themselves functions of one or more other variables. Suppose, for example, that x and y are functions of t, so that a small change δt in t produces small changes δx in x and δy in y. The change δz in $z = f(x, y)$ is then

$$\delta z = f(x + \delta x, y + \delta y) - f(x, y)$$
$$= f(x + \delta x, y + \delta y) - f(x, y + \delta y) + f(x, y + \delta y) - f(x, y)$$

Hence,

$$\frac{\delta z}{\delta t} = \frac{f(x + \delta x, y + \delta y) - f(x, y + \delta y)}{\delta t} + \frac{f(x, y + \delta y) - f(x, y)}{\delta t}$$

$$= \frac{f(x+\delta x, y+\delta y) - f(x, y+\delta y)}{\delta x}\frac{\delta x}{\delta t} + \frac{f(x, y+\delta y) - f(x, y)}{\delta y}\frac{\delta y}{\delta t}$$

Letting $\delta t \to 0$ and assuming that all limits exist, we obtain, with the use of the above definitions,

$$\frac{df}{dt} = \frac{dz}{dt} = f_x\frac{dx}{dt} + f_y\frac{dy}{dt}$$

which can be written in the alternative form

$$\frac{dz}{dt} = \frac{\partial z}{\partial x}\frac{dx}{dt} + \frac{\partial z}{\partial y}\frac{dy}{dt}$$

This is one example of a chain rule; notice that a d is used when the derivative is of a function of one variable only; otherwise ∂ is used.

We list some other chain rules, which may be derived in a similar way to the above one.

First, if $w = f(x, y, z)$ and x, y and z are each functions of t, then the appropriate chain rule is

$$\frac{dw}{dt} = \frac{\partial w}{\partial x}\frac{dx}{dt} + \frac{\partial w}{\partial y}\frac{dy}{dt} + \frac{\partial w}{\partial z}\frac{dz}{dt}$$

If $z = f(x, y)$ and x and y are each functions of *two* variables u and v, then f varies with u and v, so that f has a partial derivative with respect to both u and v. These are given by

$$\frac{\partial z}{\partial u} = \frac{\partial z}{\partial x}\frac{\partial x}{\partial u} + \frac{\partial z}{\partial y}\frac{\partial y}{\partial u} \quad \text{and} \quad \frac{\partial z}{\partial v} = \frac{\partial z}{\partial x}\frac{\partial x}{\partial v} + \frac{\partial z}{\partial y}\frac{\partial y}{\partial v}$$

These rules can easily be recalled. In the last, for example, we know that z varies with u and v, so that it must have partial derivatives with respect to both u and v. z also varies with x and y, so its partial derivatives with respect to u and v must contain terms in both $\partial z/\partial x$ and $\partial z/\partial y$; we can therefore write, for example, for $\partial z/\partial u$

$$\frac{\partial z}{\partial u} = a\frac{\partial z}{\partial x} + b\frac{\partial z}{\partial y}$$

where a and b are multipliers. We now choose $a = \partial x/\partial u$ so that 'the ∂x in its numerator cancels the denominator ∂x of $\partial z/\partial x$ and leaves the fraction $\partial z/\partial u$', which is the same as the left-hand side. The quotation marks are to emphasise that what we have done is not mathematically valid, since the quantities involved cannot be treated as fractions which we can cancel. However, the idea of cancellation does provide a convenient way of recalling the rule. Similarly, we choose $b = \partial y/\partial u$.

EXERCISES 6.6.1

1 What are the chain rules for $\partial w/\partial u$ and $\partial w/\partial v$ in the situation where w is a function of x, y and z, and x, y and z are functions of u and v?

2 Let $x = 2uv$, $y = u^2 - v^2$ and f be a function of x and y. Show that

$$\frac{\partial f}{\partial u} = 2v\frac{\partial f}{\partial x} + 2u\frac{\partial f}{\partial y}$$

6.7 DIRECTIONAL DERIVATIVES

We defined the partial derivative f_x of a function f with respect to x as the slope of the cross-section of f in a plane $y = $ constant. We might refer to this as the derivative of f in the direction of x, since, as y is kept constant, the only change that can take place is in the direction of x. Similarly, f_y may be regarded as the derivative of f in the direction of y. f_x and f_y are examples of *directional derivatives*. We now consider the slope of cross-sections of f in other planes. The plane which is at a fixed angle θ to the xz plane and goes through the line of intersection of the planes $x = x'$ and $y = y'$ is given parametrically by $x = x' + t\cos\theta$, $y = y' + t\sin\theta$; part of it is shown shaded in Figure 6.6, together with its curve of intersection with f. Putting these expressions for x and y into $f(x, y)$ gives $f(x' + t\cos\theta, y' + t\sin\theta)$, so that, since x', y' and θ are all fixed, f is a function of t. Differentiating f with respect to t will give us the slope of the cross-section of f in this plane, that is, the derivative of f in the direction $(\cos\theta, \sin\theta)$. We use the chain rule for $z = f(x, y)$ with x and y functions of t, which is

$$\frac{dz}{dt} = \frac{\partial z}{\partial x}\frac{dx}{dt} + \frac{\partial z}{\partial y}\frac{dy}{dt}$$

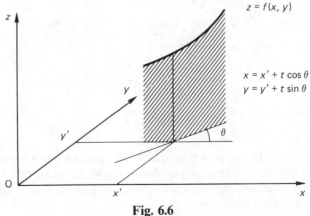

Fig. 6.6

After performing the differentiation, we evaluate at $t = 0$, to obtain the directional derivative of f at the point (x', y') in the direction $(\cos \theta, \sin \theta)$ as

$$f_x(x', y') \cos \theta + f_y(x', y') \sin \theta$$

Example 6.7.1

Let $f(x, y) = x^2 + y^2$. Then $f_x = 2x$ and $f_y = 2y$ and the directional derivative at $(1, -2, 5)$ is given by $2 \cos \theta - 4 \sin \theta$.

EXERCISE 6.7.1

Find the directional derivatives of:

(i) $f(x, y) = x^2 - y^2$ at $(1, 1, 0)$ in the direction $\theta = \pi/4$;
(ii) $f(x, y) = xy$ at $(2, 2, 4)$ in the direction $\theta = \pi/3$;
(iii) $f(x, y) = 1/(x + y^2)$ at $(0, 1, 1)$ in the direction $\theta = \pi/6$.

6.8 HIGHER PARTIAL DERIVATIVES

Let f be a function of x and y. Then, as we have seen, the partial derivatives f_x and f_y are also functions of x and y, and so must also possess partial derivatives with respect to x and y. With $z = f(x, y)$, we obtain the *second partial derivatives* by partially differentiating $\dfrac{\partial z}{\partial x} = f_x$ and $\dfrac{\partial z}{\partial y} = f_y$ with respect to x and y as

$$\frac{\partial}{\partial x}\left(\frac{\partial z}{\partial x}\right) = \frac{\partial^2 z}{\partial x^2} = (z_x)_x = z_{xx} = \frac{\partial^2 f}{\partial x^2} = f_{xx}$$

$$\frac{\partial}{\partial y}\left(\frac{\partial z}{\partial x}\right) = \frac{\partial^2 z}{\partial y \partial x} = (z_x)_y = z_{xy} = \frac{\partial^2 f}{\partial y \partial x} = f_{xy}$$

$$\frac{\partial}{\partial x}\left(\frac{\partial z}{\partial y}\right) = \frac{\partial^2 z}{\partial x \partial y} = (z_y)_x = z_{yx} = \frac{\partial^2 f}{\partial x \partial y} = f_{yx}$$

$$\frac{\partial}{\partial y}\left(\frac{\partial z}{\partial y}\right) = \frac{\partial^2 z}{\partial y^2} = (z_y)_y = z_{yy} = \frac{\partial^2 f}{\partial y^2} = f_{yy}$$

Notice particularly the order in which the variables are written. In the subscript form of f_{xy}, for example, the order of the subscripts is the same as the order of differentiating, that is, with respect to x first. In the other notation, the order is reversed.

Example 6.8.1

Find the second partial derivatives of

$$z = x \sin y + \cos xy$$

$z_x = \sin y - y \sin xy$ and $z_y = x \cos y - x \sin xy$
Hence,

$$z_{xx} = \frac{\partial}{\partial x} z_x = -y^2 \cos xy$$

$$z_{xy} = \frac{\partial}{\partial y} z_x = \cos y - \sin xy - xy \cos xy$$

$$z_{yx} = \frac{\partial}{\partial x} z_y = \cos y - \sin xy - xy \cos xy$$

$$z_{yy} = \frac{\partial}{\partial y} z_y = -x \sin y - x^2 \cos xy$$

We notice in this example that $z_{xy} = z_{yx}$; although this is not universally true, it is true for functions which are 'well-behaved' in some sense. In particular, all the functions in our standard family possess this property. In this text, every function f, even if (like the one in Example 6.8.2) it is not given explicitly, will be assumed to satisfy $f_{xy} = f_{yx}$.

The above notation extends naturally to functions of more variables (see section-end exercises). Let us now see what happens to a chain rule under repeated differentiation. We first write the chain rule as

$$\frac{\partial z}{\partial u} = \frac{\partial x}{\partial u} \frac{\partial z}{\partial x} + \frac{\partial y}{\partial u} \frac{\partial z}{\partial y}$$

We interpret this in the following way: to differentiate z partially with respect to u, we differentiate z with respect to x and multiply by $\partial x/\partial u$, differentiate z with respect to y and multiply by $\partial y/\partial u$ and add. We write this in *operator form*

$$\frac{\partial}{\partial u}(z) = \left(\frac{\partial x}{\partial u} \frac{\partial}{\partial x} + \frac{\partial y}{\partial u} \frac{\partial}{\partial y} \right) z$$

which simply means that the *operators*

$$\frac{\partial}{\partial u} \quad \text{and} \quad \frac{\partial x}{\partial u} \frac{\partial}{\partial x} + \frac{\partial y}{\partial u} \frac{\partial}{\partial y}$$

give the same result when operating on z or any other function. We shall say that the operators are equal and use them interchangeably as the need arises.

Example 6.8.2

Find f_{uu}, f_{uv} and f_{vv} when f is a function of x and y and $x = 2uv$, $y = u^2 - v^2$.
First,

$$\frac{\partial f}{\partial u} = \frac{\partial x}{\partial u}\frac{\partial f}{\partial x} + \frac{\partial y}{\partial u}\frac{\partial f}{\partial y}$$

$$= 2v\frac{\partial f}{\partial x} + 2u\frac{\partial f}{\partial y}$$

$$= \left(2v\frac{\partial}{\partial x} + 2u\frac{\partial}{\partial y} \right)f$$

so the operators satisfy the equation

$$\frac{\partial}{\partial u} = 2v\frac{\partial}{\partial x} + 2u\frac{\partial}{\partial y} \tag{1}$$

Similarly, the chain rule

$$\frac{\partial f}{\partial v} = \frac{\partial x}{\partial v}\frac{\partial f}{\partial x} + \frac{\partial y}{\partial v}\frac{\partial f}{\partial y}$$

yields

$$\frac{\partial f}{\partial v} = 2u\frac{\partial f}{\partial x} - 2v\frac{\partial f}{\partial y}$$

and

$$\frac{\partial}{\partial v} = 2u\frac{\partial}{\partial x} - 2v\frac{\partial}{\partial y} \tag{2}$$

Now

$$f_{uu} = \frac{\partial}{\partial u} f_u$$

$$= \frac{\partial}{\partial u}\left(2v\frac{\partial f}{\partial x} + 2u\frac{\partial f}{\partial y} \right)$$

$$= 2v\frac{\partial}{\partial u}\frac{\partial f}{\partial x} + 2\frac{\partial f}{\partial y} + 2u\frac{\partial}{\partial u}\frac{\partial f}{\partial y}$$

since v is constant when we differentiate with respect to u and we have
used the product rule on the second term. Now replace $\partial/\partial u$ by the

right-hand operator in Equation (1), to obtain

$$f_{uu} = 2v\left(2v\frac{\partial}{\partial x} + 2u\frac{\partial}{\partial y}\right)\frac{\partial f}{\partial x} + 2\frac{\partial f}{\partial y} + 2u\left(2v\frac{\partial}{\partial x} + 2u\frac{\partial}{\partial y}\right)\frac{\partial f}{\partial y}$$

$$= 4v^2 f_{xx} + 4uv f_{xy} + 2f_y + 4uv f_{yx} + 4u^2 f_{yy}$$

$$= 4v^2 f_{xx} + 8uv f_{xy} + 4u^2 f_{yy} + 2f_y$$

Similarly, we find

$$f_{vv} = 4u^2 f_{xx} - 8uv f_{xy} + 4v^2 f_{yy} - 2f_y$$

To find f_{uv}, we can differentiate in either order: for example, differentiating with respect to u first,

$$f_{uv} = \frac{\partial}{\partial v} f_u$$

$$= \frac{\partial}{\partial v}(2vf_x + 2uf_y)$$

$$= 2f_x + 2v\frac{\partial}{\partial v} f_x + 2u\frac{\partial}{\partial v} f_y$$

$$= 2f_x + 2v\left(2u\frac{\partial}{\partial x} - 2v\frac{\partial}{\partial y}\right)f_x + 2u\left(2u\frac{\partial}{\partial x} - 2v\frac{\partial}{\partial y}\right)f_y \quad \text{by Equation (2)}$$

$$= 2f_x + 4uv f_{xx} - 4v^2 f_{xy} + 4u^2 f_{xy} - 4uv f_{yy}$$

EXERCISES 6.8.1

1 Find the second partial derivatives of $z = x \cosh xy$.

2 Find w_{xyz} when $w = \ln(x + y^2 + z^3)$.

3 A function f is said to satisfy Laplace's equation if $f_{xx} + f_{yy} = 0$. Show that functions given as follows satisfy Laplace's equation:

(i) $f(x, y) = 4x^3 - 12xy^2$; (ii) $f(x, y) = e^x \cos y$;
(iii) $f(x, y) = \tan^{-1}(y/x)$.

4 Show that when f is a function of x and y, and $x = 2uv$, $y = u^2 - v^2$, and that

$$f_{uu} + f_{vv} = 4(u^2 + v^2)(f_{xx} + f_{yy})$$

5 Derive the formula

$$f_{uv} = f_{xx} x_u x_v + f_{yy} y_u y_v + f_{xy}(x_u y_v + x_v y_u) + f_x x_{uv} + f_y y_{uv}$$

Find similar formulae for f_{uu} and f_{vv}.

6 Let f be a function of x and y and let $x = r\cos\theta$, $y = r\sin\theta$. Express f_{rr} and $f_{\theta\theta}$ in terms of r, θ and partial derivatives of f with respect to x and y. Deduce that

$$f_{rr} + \frac{1}{r^2}f_{\theta\theta} + \frac{1}{r}f_r = f_{xx} + f_{yy}$$

6.9 MAXIMA AND MINIMA

Let f be a function of two variables. We shall say that f has a *local maximum* at the point (x', y') if $f(x' + h, y' + k) < f(x', y')$ for all sufficiently small but non-zero values of h and k. We need to specify 'sufficiently small' values of h and k because there may be functions which decrease initially as we move away from (x', y'), but shortly afterwards increase to values greater than $f(x', y')$; this also explains the need to include the word 'local' in the description. A *global* maximum occurs at the point (x', y') if $f(x', y') > f(x, y)$ for *every* point (x, y) in the domain of f. The definitions of local and global minima are obtained from those of the corresponding maxima simply by reversing the inequalities.

If f has a local maximum or minimum at (x', y'), then the tangents at this point to its cross-sections in all planes through the line of intersection of $x = x'$ and $y = y'$ must be parallel to the xy plane. A local minimum is illustrated in Figure 6.7. But the slope of the cross-section of f in the plane $x = x' + t\cos\theta$, $y = y' + t\sin\theta$ is given by $f_x(x', y')\cos\theta + f_y(x', y')\sin\theta$. We must, therefore, have f_x and f_y both zero at (x', y'). A point (x', y') at which f_x and f_y both vanish is called a *stationary point*; as we shall see, f does not necessarily have a maximum or a minimum at stationary point.

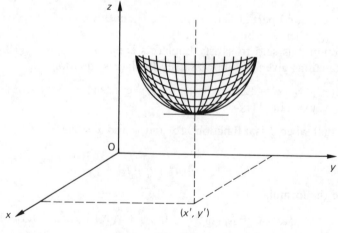

Fig. 6.7

Example 6.9.1

Describe the graphs of the following near the stationary point $(0,0)$:

(i) $z = x^2 + y^2$; (ii) $z = -x^2 - y^2$; (iii) $z = x^2 - y^2$;
(iv) $z = -x^2 + y^2$; (v) $z = x^3 + y^2$; (vi) $z = x^3 - y^2$.

We can obtain a good idea of the shape of these graphs by looking at their cross-sections in the planes $y = 0$ and $x = 0$.

In (i), these are $z = x^2$ and $z = y^2$, respectively, and the point $(0,0)$ corresponds to a local minimum. Similarly, (ii) has a local maximum at $(0,0)$.

In (iii), however, $z = x^2$ and $z = -y^2$ are parabolas pointing in the opposite directions, so that $(0,0)$ is no longer a maximum or a minimum; we call it a *saddle point* for obvious reasons. (iv) is the upside-down version of (iii).

In (v) and (vi), the cross-section of f in the plane $y = 0$ is $z = x^3$, and the resulting graphs could be described as an armchair and a shelf (or alp), respectively. These results are shown schematically in Table 6.9.1, in which the reader is supposed to be looking down on the xy plane; the stationary point is depicted as an asterisk.

Table 6.9.1

up	down	down
up * up	down * down	up * up
up	down	down
(a) minimum	(b) maximum	(c) saddle point
up	up	down
down * down	down * up	down * up
up	up	down
(d) saddle point	(e) armchair	(f) shelf

We now proceed to find conditions on f which will enable us to decide upon the nature of a stationary point. Let the function g be defined by

$$g(t) = f(x' + t \cos \theta, y' + t \sin \theta)$$

Then $z = g(t)$ is the cross-section of f in the plane $x = x' + t \cos \theta$, $y = y' + t \sin \theta$. For each value of θ, we use the ordinary one-variable conditions for g to have a maximum or minimum at $t = 0$ to obtain criteria for f to have a maximum, minimum or saddle point at (x', y'). We shall assume that f is sufficiently well-behaved for f_{xy} to equal f_{yx}.

We have $g(t) = f(x, y)$, where $x = x' + t \cos \theta$, $y = y' + t \sin \theta$. The chain rule gives

$$\frac{dg}{dt} = \frac{\partial f}{\partial x} \cos \theta + \frac{\partial f}{\partial y} \sin \theta \tag{1}$$

We now differentiate with respect to t by replacing d/dt by

$$\cos \theta \frac{\partial}{\partial x} + \sin \theta \frac{\partial}{\partial y}$$

to obtain

$$\frac{d^2 g}{dt^2} = \left(\cos \theta \frac{\partial}{\partial x} + \sin \theta \frac{\partial}{\partial y} \right)(f_x \cos \theta + f_y \sin \theta)$$

$$= \cos^2 \theta f_{xx} + 2 \cos \theta \sin \theta f_{xy} + \sin^2 \theta f_{yy} \tag{2}$$

If $dg/dt = 0$ at $t = 0$, then g has a maximum at $t = 0$ if d^2g/dt^2 is negative or a minimum if it is positive. Thus, if d^2g/dt^2 is negative at $t = 0$ for all angles θ, then f will have a local maximum at (x', y'), while if d^2g/dt^2 is positive for all values of θ, then f will have a local minimum there. If, however, there are some values of θ for which d^2g/dt^2 is positive and some for which it is negative, then f has a saddle point. The quantity which distinguishes between these cases is the *discriminant*,

$$\Delta = f_{xx} f_{yy} - f_{xy}^2$$

where the derivatives are evaluated at $t = 0$, that is, at the point (x', y').

Case I, $\Delta > 0$:   Here we must have $f_{xx} f_{yy} > 0$ so that $f_{xx} \neq 0$ and we can write

$$\frac{d^2 g}{dt^2} = \frac{1}{f_{xx}}((f_{xx} \cos \theta + f_{xy} \sin \theta)^2 + \Delta \sin^2 \theta) \tag{3}$$

The reader should check that this is indeed the same as the right-hand side of Equation (2). It is now clear that d^2g/dt^2 has the same sign as f_{xx}, so that the stationary point is a local maximum if $f_{xx} < 0$ or a local minimum if $f_{xx} > 0$.

Case II, $\Delta < 0$:   If $f_{xx} \neq 0$, Equation (3) still holds but the sign now depends on the value of θ; for example, if $\theta = 0$, the sign of d^2g/dt^2 is the same as that of f_{xx}, while if $\theta = \tan^{-1}(-f_{xx}/f_{yy})$, it takes the opposite sign. In this case, we must have a saddle point.

If $f_{xx} = 0$, then $f_{xy} \neq 0$ and Equation (2) gives

$$\frac{d^2 g}{dt^2} = \sin^2 \theta (2 f_{xy} \cot \theta + f_{yy}),$$

so that the right-hand side takes both positive and negative values, thus giving a saddle point again.

If the discriminant turns out to be zero at (x', y'), further investigation would be required to determine the nature of the stationary point, but this is beyond our scope.

Table 6.9.2 summarizes the situation for a function f for which $f_{xy} = f_{yx}$.

Table 6.9.2

f_x	f_y	$f_{xx}f_{yy} - f_{xy}^2$	f_{xx}	Then f has a:
0	0	>0	<0	local maximum
0	0	>0	>0	local minimum
0	0	<0		saddle point
0	0	$=0$		unknown

Example 6.9.2

Find and classify the stationary points of the function f given by

$$f(x, y) = x^3 + 3y^3 - \tfrac{1}{2}x^2 - 2x - 9y$$

We have

$$f_x = 3x^2 - x - 2 = (x - 1)(3x + 2)$$
$$f_y = 9y^2 - 9 = 9(y - 1)(y + 1)$$

The stationary points occur at all points (x, y) which make f_x and f_y both zero. These are $(1, 1)$, $(1, -1)$, $(-\tfrac{2}{3}, 1)$ and $(-\tfrac{2}{3}, -1)$. We also find $f_{xx} = 6x - 1$, $f_{xy} = 0$ and $f_{yy} = 18y$, so that $\Delta = 18y(6x - 1)$. We summarise the results in Table 6.9.3.

Table 6.9.3

Point	f_{xx}	Δ	so:
$(1, 1)$	5	90	local minimum
$(1, -1)$	5	-90	saddle point
$(-2/3, 1)$	-5	-90	saddle point
$(-2/3, -1)$	-5	90	local maximum

EXERCISE 6.9.1

Find and classify the stationary points of the functions given by

(i) $f(x, y) = (x^2 + y^2)e^{-(x+y)}$;
(ii) $f(x, y) = x(1 + y^2)/2 - \tan^{-1} x$;
(iii) $f(x, y) = (x - y)^3 + x^2 y^2$.

MISCELLANEOUS EXERCISES 6

Omitting division by δt in the derivation of the chain rule in Section 6.6 gives

$$\delta z = \frac{f(x + \delta x, y + \delta y) - f(x, y + \delta y)}{\delta x}\delta x + \frac{f(x, y + \delta y) - f(x, y)}{\delta y}\delta y$$

$$\approx \frac{\partial f}{\partial x}\delta x + \frac{\partial f}{\partial y}\delta y \text{ for small } \delta x, \delta y$$

$$= \frac{\partial z}{\partial x}\delta x + \frac{\partial z}{\partial y}\delta y$$

Here δz is the *total change* in z caused by changes $\delta x, \delta y$ in x, y. This type of formula enables us to investigate the effect on the answer of small changes in the data. The first two exercises illustrate this.

1 The area S of a triangle is given by the formula $S = \frac{1}{2}bc \sin A$, where A is the angle between two sides of the triangle of length b, c. Find an approximation to the change δS in S caused by small changes $\delta b, \delta c, \delta A$ in b, c, A, and show that

$$\frac{\delta S}{S} \approx \frac{\delta b}{b} + \frac{\delta c}{c} + A \cot A \frac{\delta A}{A}$$

Show that if $A = \pi/4$, then the maximum percentage error in S due to maximum errors of 1% in b, c, A is about 2.8%.

2 Suppose that in calculating the period $2\pi\sqrt{l/g}$ of a simple pendulum of length l the value $22/7$ is used for π and 9.80 m/s^2 is used for g (instead of 9.81 m/s^2). Find the percentage error in the period.

3 With $x = r \cos \theta$, $y = r \sin \theta$, write down expressions for $\partial x/\partial r$ and $\partial y/\partial \theta$. Now express r, θ in terms of x, y and hence find $\partial r/\partial x, \partial \theta/\partial y$. Show that $\partial x/\partial r = \partial r/\partial x$, but that $\partial y/\partial \theta \neq \partial \theta/\partial y$.

4 The pressure (p) of a gas is given in terms of its volume (V) and temperature (T) by the equation $p = RT/V$, where R is the gas constant. Find $\partial p/\partial V$. Find V in terms of p, T and hence $\partial V/\partial T$; similarly find

$\partial T/\partial p$ and show that

$$\frac{\partial p}{\partial V}\frac{\partial V}{\partial T}\frac{\partial T}{\partial p} = -1$$

5 Repeat the last question when the pressure, volume and temperature of the gas are connected by van der Waals' equation

$$p = \frac{RT}{V-b} - \frac{a}{V^2},$$

where a, b are constants. (*Hint*: To evaluate $\partial V/\partial T$ differentiate van der Waals' equation implicitly with respect to T, keeping p fixed.)

6 Show that the wave equation

$$\frac{\partial^2 y}{\partial x^2} = \frac{1}{c^2}\frac{\partial^2 y}{\partial t^2}$$

where c is the constant speed of the wave, is satisfied by $y = A\sin(r\pi x/l)\sin(r\pi ct/l)$ for arbitrary constants A, r, l.

7 Show that the diffusion equation

$$\frac{\partial V}{\partial t} = k\frac{\partial^2 V}{\partial x^2}$$

where k is a constant, is satisfied by $V = (A\cos rx + B\sin rx)e^{-kr^2 t}$, where A, B, r are arbitrary constants.

8 Find the directional derivatives at the origin in the direction $(\cos\theta, \sin\theta)$ of the functions given by

(i) $f(x, y) = 3x + 4y$;
(ii) $f(x, y) = \sqrt{x + y + 4}$;
(iii) $f(x, y) = (y + 2)\ln(x + 1)$.

In each case find the maximum value of the directional derivative at the origin and the value of θ for which it occurs.

9 Find the maximum value of $(xyz)^2$ subject to the constraint $x^2 + y^2 + z^2 = 3$. (*Hint*: Eliminate one of the variables using the constraint.)

6.10 ANSWERS TO EXERCISES

Exercise 6.2.1

(i) The biggest domain is the whole of the xy plane except the point $(0, 0)$, the range is $\mathbb{R}^+$; (ii) the biggest domain is the whole of the xy plane excluding points on the unit circle, the range is $(-\infty, -1] \cup (0, \infty)$; (iii) the biggest domain is the set of all points of the xy plane for which $x - y > 0$, the range is $\mathbb{R}$.

Exercises 6.3.1

1 (i) The straight line $z = 1 + y$ in the plane $x = 1$, (ii) the curve $z = x^3 + 1$ in the plane $y = 1$ (this is a cubic), (iii) $f(t/\sqrt{2}, t/\sqrt{2}) = 0$, so the cross-section is the line $y = x$ in the xy plane. (iv) $f(t/2, t\sqrt{3}/2) = -t^2/2$, which is a parabola.

2 The level curves are circles $x^2 + y^2 = k$ in the planes $z = k$. (The cross-sections in planes $x = k$, $y = k$ are parabolas $z = k + y^2$, $z = x^2 + k$.)

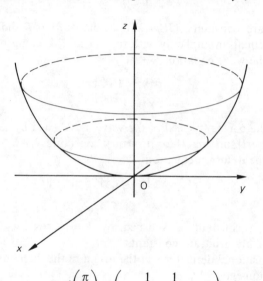

3
$$\mathbf{r}'\left(\frac{\pi}{4}\right) = \left(-\frac{1}{\sqrt{2}}, \frac{1}{\sqrt{2}}, -2\right)$$

and this is a tangent vector at

$$t = \frac{\pi}{4}.$$

$\Bigg($ The equation of the tangent line is

$$\mathbf{r} = \left(\frac{1}{\sqrt{2}}, \frac{1}{\sqrt{2}}, 0\right) + \lambda\left(-\frac{1}{\sqrt{2}}, \frac{1}{\sqrt{2}}, -2\right).\Bigg)$$

4 For points on the given curve $x^2 + y^2 = t^2 \cos^2 t + t^2 \sin^2 t = t^2 = z$, so the curve lies on the surface of the paraboloid $z = x^2 + y^2$. $(t \cos t, t \sin t)$, $t \geq 0$ is a spiral in the plane, so $(t \cos t, t \sin t, t^2)$, $t \geq 0$ is a *parabolic helix*, spiralling on the paraboloid $z = x^2 + y^2$.

5 For points on the intersection of the surface and the cylinder, $z = x^2 - y^2 = \cosh^2 t - \sinh^2 t = 1$, so the curve of intersection is the hyperbola $x^2 - y^2 = 1$ in the plane $z = 1$ for $x > 0$.

Exercises 6.5.1

1 (i) $f_x(x, y) = 2x$, $f_y(x, y) = 2y$, (ii) $f_x(x, y) = f_y(x, y) = -\dfrac{1}{(x + y)^2}$,

(iii) $f_x(x, y) = \dfrac{2x}{x^2 + y^2 + 1}$, $f_y(x, y) = \dfrac{2y}{x^2 + y^2 + 1}$,

(iv) $f_x(x, y) = \dfrac{1}{1 + (y/x)^2} \dfrac{-y}{x^2} = \dfrac{-y}{x^2 + y^2}$, $f_y(x, y) = \dfrac{1}{1 + (y/x)^2} \dfrac{1}{x}$

$$= \dfrac{x}{x^2 + y^2},$$

(v) $f_x(x, y) = (1 + xy^2)e^{xy^2}$, $f_y(x, y) = 2x^2 y e^{xy^2}$,

(vi) $f_x(x, y) = yx^{y-1}$, $f_y(x, y) = x^y \ln x$.

2 (i) $f_x(x, y, z) = z \cos(yz^3 + x)$, $f_y(x, y, z) = z^4 \cos(yz^3 + x)$,
$f_z(x, y, z) = \sin(yz^3 + x) + 3yz^3 \cos(yz^3 + x)$,

(ii) $f_x(x, y, z) = \dfrac{1}{x}$, $f_y(x, y, z) = \dfrac{2}{y}$, $f_z(x, y, z) = \dfrac{3}{z}$.

Exercises 6.6.1

1 We know that $\partial w/\partial u$ and $\partial w/\partial v$ must contain terms in $\partial w/\partial x$, $\partial w/\partial y$ and $\partial w/\partial z$. Thus

$$\frac{\partial w}{\partial u} = \frac{\partial w}{\partial x}\frac{\partial x}{\partial u} + \frac{\partial w}{\partial y}\frac{\partial y}{\partial u} + \frac{\partial w}{\partial z}\frac{\partial z}{\partial u}$$

and

$$\frac{\partial w}{\partial v} = \frac{\partial w}{\partial x}\frac{\partial x}{\partial v} + \frac{\partial w}{\partial y}\frac{\partial y}{\partial v} + \frac{\partial w}{\partial z}\frac{\partial z}{\partial v}.$$

2
$$\frac{\partial f}{\partial u} = \frac{\partial f}{\partial x}\frac{\partial x}{\partial u} + \frac{\partial f}{\partial y}\frac{\partial y}{\partial u} = 2v\frac{\partial f}{\partial x} + 2u\frac{\partial f}{\partial y}.$$

Exercise 6.7.1

(i) $f_x(1, 1)\cos\dfrac{\pi}{4} + f_y(1, 1)\sin\dfrac{\pi}{4} = 2 \cdot \dfrac{1}{\sqrt{2}} - 2 \cdot \dfrac{1}{\sqrt{2}} = 0$;

(ii) $2 \cdot \dfrac{1}{2} + 2 \cdot \dfrac{\sqrt{3}}{2} = 1 + \sqrt{3}$;

(iii) $f_x(x, y) = \dfrac{-1}{(x + y^2)^2}$, so $f_x(0, 1) = -1$.

$f_y(x, y) = \dfrac{-2y}{(x + y^2)^2}$, so $f_y(0, 1) = -2$.

The directional derivative is then

$$4 - \cos\frac{\pi}{6} - 2\sin\frac{\pi}{6} = -\frac{\sqrt{3}}{2} - 1.$$

Exercises 6.8.1

1 $z_x = \cosh xy + xy \sinh xy$, $z_y = x^2 \sinh xy$, so
$z_{xx} = 2y \sinh xy + xy^2 \cosh xy$, $z_{yx} = 2x \sinh xy + x^2 y \cosh xy$,
$z_{xy} = 2x \sinh xy + x^2 y \cosh xy$, $z_{yy} = x^3 \cosh xy$.

2 $w_x = \dfrac{1}{x + y^2 + z^3}$, $w_{xy} = \dfrac{-2y}{(x + y^2 + z^3)^2}$, $w_{xyz} = \dfrac{12yz^2}{(x + y^2 + z^3)^3}$.

3 (i) $f_{xx} + f_{yy} = 24x - 24x = 0$, (ii) $f_{xx} + f_{yy} = e^x \cos y - e^x \cos y = 0$,

(iii) $f_x = \dfrac{-y}{x^2 + y^2}$, $f_{xx} = \dfrac{2xy}{(x^2 + y^2)^2}$, $f_y = \dfrac{x}{x^2 + y^2}$, $f_{yy} = \dfrac{-2xy}{(x^2 + y^2)^2}$,

so $f_{xx} + f_{yy} = 0$.

4 $f_{uu} = 4v^2 f_{xx} + 8uv f_{xy} + 4u^2 f_{yy} + 2f_y$ from Example 6.8.2. Using expressions for the partial differential operators from this example,

$$f_{vv} = \frac{\partial}{\partial v} f_v = \frac{\partial}{\partial v}(2uf_x - 2vf_y) = 2u\frac{\partial}{\partial v} f_x - 2v\frac{\partial}{\partial v} f_y - 2f_y$$

$$= 2u\left(2u\frac{\partial}{\partial x} - 2v\frac{\partial}{\partial y}\right) f_x - 2v\left(2u\frac{\partial}{\partial x} - 2v\frac{\partial}{\partial y}\right) f_y - 2f_y$$

$$= 4u^2 f_{xx} - 8uv f_{xy} + 4v^2 f_{yy} - 2f_y.$$

The answer follows upon adding f_{uu} and f_{vv}.

5 $f_u = f_x x_u + f_y y_u$, so $f_{uv} = (f_x x_u + f_y y_u)_v$

$$= (f_x)_v x_u + f_x x_{uv} + (f_y)_v y_u + f_y y_{uv}$$

$$= (f_{xx} x_v + f_{xy} y_v) x_u + f_x x_{uv} + (f_{yx} x_v + f_{yy} y_v) y_u + f_y y_{uv},$$

which simplifies to the given answer.

6 $\qquad\qquad f_r = f_x x_r + f_y y_r = \cos\theta f_x + \sin\theta f_y,$

so

$$\frac{\partial}{\partial r} = \cos\theta\frac{\partial}{\partial x} + \sin\theta\frac{\partial}{\partial y}.$$

Also,

$$f_\theta = f_x x_\theta + f_y y_\theta = -r \sin \theta f_x + r \cos \theta f_y,$$

so

$$\frac{\partial}{\partial \theta} = -r \sin \theta \frac{\partial}{\partial x} + r \cos \theta \frac{\partial}{\partial y}.$$

$$f_{rr} = \left(\cos \theta \frac{\partial}{\partial x} + \sin \theta \frac{\partial}{\partial y} \right) (\cos \theta f_x + \sin \theta f_y)$$

$$= \cos^2 \theta f_{xx} + 2 \cos \theta \sin \theta f_{xy} + \sin^2 \theta f_{yy},$$

$$f_{\theta\theta} = -r \cos \theta f_x - r \sin \theta f_y - r \sin \theta \left(-r \sin \theta \frac{\partial}{\partial x} + r \cos \theta \frac{\partial}{\partial y} \right) f_x$$

$$+ r \cos \theta \left(-r \sin \theta \frac{\partial}{\partial x} + r \cos \theta \frac{\partial}{\partial y} \right) f_y = r^2 \sin^2 \theta f_{xx}$$

$$- 2r^2 \cos \theta \sin \theta f_{xy} + r^2 \cos^2 \theta f_{yy} - r \cos \theta f_x + r \sin \theta f_y$$

Putting these into

$$f_{rr} + \frac{1}{r^2} f_{\theta\theta} + \frac{1}{r} f_r$$

gives the required result.

Exercise 6.9.1

(i) $f_x = (2x - x^2 - y^2)e^{-(x+y)}$, $f_y = (2y - x^2 - y^2)e^{-(x+y)}$

$f_x = f_y = 0 \Rightarrow x^2 + y^2 = 2x = 2y \Rightarrow x = y = 0$ or $x = y = 1$

$f_{xx} = (2 - 4x + x^2 + y^2)e^{-(x+y)}$, $f_{xy} = (-2x - 2y + x^2 + y^2)e^{-(x+y)}$,

$f_{yy} = (2 - 4y + x^2 + y^2)e^{-(x+y)}$,

so $\Delta = \{(2 - 4x + x^2 + y^2)(2 - 4y + x^2 + y^2)$

$- (-2x - 2y + x^2 + y^2)\} e^{-2(x+y)}$.

At $(0,0)$, $\Delta > 0$, $f_{xx} > 0$, so there is a local minimum, while at $(1,1)$, $\Delta < 0$, so there is a saddle point.

(ii) $f_x = \frac{1}{2}(1 + y^2) - \frac{1}{1 + x^2}$, $f_y = xy$. $f_y = 0 \Rightarrow x = 0$ or $y = 0$, $f_x = 0$ and

$x = 0 \Rightarrow y = 1$ or $y = -1$, $f_x = 0$ and $y = 0 \Rightarrow x = \pm 1$.

$$f_{xx} = \frac{2x}{(1 + x^2)^2}, \; f_{xy} = y, \; f_{yy} = x, \; \Delta = \frac{2x^2}{(1 + x^2)^2} - y^2.$$

Thus, there are saddle points at $(0, 1)$ and $(0, -1)$; at $(1, 0)$ there is a local minimum and at $(-1, 0)$ there is a local maximum.

(iii) $f_x = 3(x-y)^2 + 2xy^2$, $f_y = -3(x-y)^2 + 2x^2y$.

$f_x = f_y = 0 \Rightarrow f_x + f_y = 2xy(x+y) = 0$. The possibilities are $x = 0 = y$ or $x = -y$, which together with $f_x = 0$ gives $2x^2(6+x) = 0 \Rightarrow x = -6$.

$f_{xx} = 6(x-y) + 2y^2$, $\Delta = (6x - 6y + 2y^2)(6x - 6y + 2x^2)$ $-(-6x + 6y + 4xy)^2$.

$(-6, 6)$ is a saddle point, but $f_x = \Delta = 0$ at $(0, 0)$, so we cannot tell what happens at this point. The behaviour of the graph of the function near $(0, 0)$ is illustrated below.

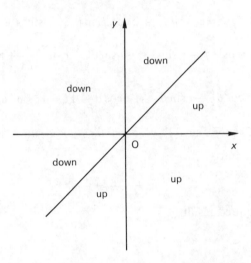

Miscellaneous Exercises 6

1 $\delta S \approx \dfrac{\partial S}{\partial b} \delta b + \dfrac{\partial S}{\partial c} \delta c + \dfrac{\partial S}{\partial A} \delta A = \dfrac{1}{2} c \sin A \delta b + \dfrac{1}{2} b \sin A \delta c + \dfrac{1}{2} bc \cos A \delta A.$

Dividing through by $S = \dfrac{1}{2} bc \sin A$ gives the required result.

$\dfrac{\delta S}{S} \approx 0.01 + 0.01 + \left(\dfrac{\pi}{4} \cot \dfrac{\pi}{4} \right) 0.01 \approx 0.028$, which corresponds to 2.8%.

2 $\delta T \approx 2 \sqrt{\dfrac{l}{g}} \, \delta \pi - \pi \sqrt{\dfrac{l}{g^3}} \, \delta g$ so $\dfrac{\delta T}{T} \approx \dfrac{\delta \pi}{\pi} - \dfrac{\delta g}{2g} \approx \dfrac{22/7 - 3.1416}{3.1416}$

$- \dfrac{9.80 - 9.81}{2 \times 9.80} \approx 0.0009$, which corresponds to 0.09%.

3 $x_r = \cos \theta$, $y_\theta = r \cos \theta$. $r = \sqrt{x^2 + y^2}$, $\theta = \tan^{-1} \dfrac{y}{x}$, so $r_x = \dfrac{x}{r} = \cos \theta = x_r$,

$\theta_y = \dfrac{1/x}{1 + y^2/x^2} = \dfrac{x}{r^2} = \dfrac{1}{r} \cos \theta \neq y_\theta.$

4 $Pv = -\dfrac{RT}{V^2}$; $V = \dfrac{RT}{p}$, $V_T = \dfrac{R}{p}$; $T = \dfrac{pV}{R}$, $T_p = \dfrac{V}{R}$, so

$$p_V V_T T_p = -\dfrac{RT}{V^2}\dfrac{R}{p}\dfrac{V}{R} = -\dfrac{RT}{pV} = -1.$$

5 $p_V = -\dfrac{RT}{(V-b)^2} + \dfrac{2a}{V^3}$; differentiating partially with respect to T,

$$0 = \dfrac{R}{V-b} - \dfrac{RT}{(V-b)^2}V_T + \dfrac{2a}{V^3}V_T \Rightarrow V_T = \dfrac{-R/(V-b)}{2a/V^3 - RT/(V-b)^2}.\text{ Solving for}$$

T, we find $T = \dfrac{\left(p + \dfrac{a}{V^2}\right)(V-b)}{R}$, so $T_p = \dfrac{V-b}{R}$. Then $p_V V_T T_p = -1$.

6 $y_{xx} = -\dfrac{r^2\pi^2}{l^2}y$, $y_{tt} = -\dfrac{r^2\pi^2 c^2}{l^2}y$, so $y_{xx} = -\dfrac{1}{c^2}y_{tt}$.

7 $V_t = -kr^2 V$, $V_{xx} = -r^2 V$, so $V_t = kV_{xx}$.

8 (i) $3\cos\theta + 4\sin\theta = 5\cos(\theta - \alpha)$, where $\alpha = \tan^{-1}\frac{4}{3}$. Maximum value 5 when $\theta = \alpha$.

(ii) $\dfrac{\cos\theta + \sin\theta}{2\sqrt{x+y+4}} = \dfrac{1}{4}(\cos\theta + \sin\theta) = \dfrac{1}{2\sqrt{2}}\cos\left(\theta - \dfrac{\pi}{4}\right)$ at $(0,0)$.

Maximum value $\dfrac{1}{2\sqrt{2}}$ at $\theta = \dfrac{\pi}{4}$.

(iii) $\dfrac{y+2}{x-1}\cos\theta + \ln(x+1)\sin\theta = 2\cos\theta$ at $(0,0)$. Maximum value 2 at $\theta = 0$.

9 $R = (xy)^2(3 - x^2 - y^2)$, so $R_x = 2xy^2(3 - x^2 - y^2) - 2x^3 y^2$
$= 2xy^2(3 - 2x^2 - y^2)$ and similarly $R_y = 2x^2 y(3 - x^2 - 2y^2)$. Setting these equal to zero, we find $x = 0$, $y = 0$, which give a minimum, or $3 - 2x^2 - y^2 = 3 - x^2 - 2y^2 = 0$ which gives $x^2 = y^2 = 1$ and hence $z^2 = 1$ to satisfy the constraint and $R = 1$.

7 LINE INTEGRALS AND DOUBLE INTEGRALS

7.1 VECTOR FIELDS

We have already met functions and vector functions of one variable, and functions of two or more variables. We now look briefly at vector functions of two or three variables. When the domain of such functions is a region of space, we usually refer to the functions as *vector fields*. Thus, a vector field is a rule which associates with each point in a region of space a vector. In this usage, we could refer to a function of two or three variables as a *scalar field*.

Example 7.1.1

Let a point mass m be placed at the origin. For every point (x, y, z) in space away from the origin let $\mathbf{v}(x, y, z)$ be the gravitational force exerted by the mass at the origin on a unit mass placed at the point (x, y, z). The magnitude of this force is

$$|\mathbf{v}(x, y, z)| = \frac{Gm}{d^2}$$

where $d^2 = x^2 + y^2 + z^2$ and G is the gravitational constant. The force acts towards the origin and a unit vector in this direction is

$$-(x\mathbf{i} + y\mathbf{j} + z\mathbf{k})/d$$

The vector field $\mathbf{v}$ is then given by

$$\mathbf{v}(x, y, z) = -\frac{Gm}{d^3}(x\mathbf{i} + y\mathbf{j} + z\mathbf{k})$$

A similar result is obtained when gravitational force is replaced by electrostatic or magnetic forces. Indeed, vector fields arise in many physical situations; another example is the flow of fluids, in which the vector field represents velocity.

Example 7.1.2

An example of a scalar field is provided by $d = \sqrt{x^2 + y^2 + z^2}$. Its gradient is given by

$$\nabla d = d_x \mathbf{i} + d_y \mathbf{j} + d_z \mathbf{k}$$

Now

$$d_x = \frac{x}{\sqrt{x^2 + y^2 + z^2}} = \frac{x}{d}$$

Similarly, $d_y = y/d$ and $d_z = z/d$. Then

$$\nabla d = \frac{(x\mathbf{i} + y\mathbf{j} + z\mathbf{k})}{d}$$

is a vector field.

EXERCISE 7.1.1

Let $f(x, y, z) = Gm/d$, where G, m and d are as in Example 7.1.1. Find ∇f and verify that it is the vector field of Example 7.1.1.

7.2 LINE INTEGRALS

The work done by a force F, whose direction is fixed and whose magnitude depends on its position, in moving an object in the direction of F from $s = s_1$ to $s = s_2$ is given by

$$W = \int_{s_1}^{s_2} F(s)\, ds$$

To generalise this, assume that we have in the xy plane a force field

$$\mathbf{F}(x, y) = P(x, y)\mathbf{i} + Q(x, y)\mathbf{j}$$

and wish to find the work done by moving an object along a smooth curve given parametrically in terms of distance s along the curve by

$$\mathbf{r}(s) = x(s)\mathbf{i} + y(s)\mathbf{j}, \qquad s_1 \leq s \leq s_2$$

We obtained the unit vector in the direction of the tangent to the curve in Section 5.11 as $d\mathbf{r}/ds$, so the component of the force along the curve is $\mathbf{F} \cdot d\mathbf{r}/ds$ and the required work is

$$W = \int_{s_1}^{s_2} \left(\mathbf{F} \cdot \frac{d\mathbf{r}}{ds} \right) ds$$

$$= \int_{s_1}^{s_2} \left(P(x(s), y(s))\mathbf{i} + Q(x(s), y(s))\mathbf{j} \right) \cdot \left(\frac{dx}{ds}\mathbf{i} + \frac{dy}{ds}\mathbf{j} \right) ds$$

$$= \int_{s_1}^{s_2} \left(P(x(s), y(s))\frac{dx}{ds} + Q(x(s), y(s))\frac{dy}{ds} \right) ds$$

By changing the variable of integration to x in the first integral and y in the second, we obtain the integral in the simple form

$$W = \int_C (P(x, y)\, dx + Q(x, y)\, dy) \tag{1}$$

where C is the curve along which we integrate. An integral of this form is called a *line integral*. We can express it more neatly as

$$W = \int_C \mathbf{F} \cdot d\mathbf{r}$$

where $d\mathbf{r} = \mathbf{i}\, dx + \mathbf{j}\, dy$.

If C has equation $y = f(x)$, then we can evaluate the integral by substituting for y in each term of Equation (1) to obtain an ordinary integral

$$W = \int_{x_1}^{x_2} (P(x, f(x)) + Q(x, f(x))f'(x))\, dx$$

where x_1 and x_2 are the values of x at the beginning and end of C.

If the curve is specified in terms of a parameter t, we can change the variables of integration in Equation (1) into t, to obtain

$$W = \int_{t_1}^{t_2} \left(P(x(t), y(t))\frac{dx}{dt} + Q(x(t), y(t))\frac{dy}{dt} \right) dt \tag{2}$$

where t_1 and t_2 are the values of t corresponding to s values of s_1 and s_2 respectively.

Example 7.2.1

Evaluate the line integral

$$I = \int_C (x^2 - y)\,dx + (y^2 + x)\,dy$$

where C is the curve given by $y = x^2 + 1$, $0 \le x \le 1$.

Replacing y by $x^2 + 1$ and dy by $2x\,dx$, we obtain for the integral

$$I = \int_0^1 ((x^2 - x^2 - 1) + 2x((x^2 + 1)^2 + x))\,dx$$

$$= \int_0^1 (2x^5 + 4x^3 + 2x + 2x^2 - 1)\,dx$$

$$= 2$$

Example 7.2.2

Evaluate the line integral $\int_C (x^2 + y)\,dx + (x + y^2)\,dy$, where C is the curve given by $x = t$, $y = t + 1$, $0 \le t \le 2$.

Since C is given parametrically, we shall use the form (2). We have $x'(t) = y'(t) = 1$, so we obtain for the integral

$$\int_C (x^2 + 2y)\,dx + (x + y^2)\,dy = \int_0^2 (t^2 + 2t + 2 + t + (t + 1)^2)\,dt$$

$$= \frac{64}{3}$$

If we have a vector field in space,

$$\mathbf{v}(x, y, z) = P(x, y, z)\mathbf{i} + Q(x, y, z)\mathbf{j} + R(x, y, z)\mathbf{k}$$

and a curve C in space, then we can define the line integral

$$\int_C \mathbf{v} \cdot d\mathbf{r} = \int_C P(x, y, z)\,dx + Q(x, y, z)\,dy + R(x, y, z)\,dz$$

where in this case $d\mathbf{r} = \mathbf{i}\,dx + \mathbf{j}\,dy + \mathbf{k}\,dz$.

205

Example 7.2.3

Find $\int_C y \, dx - x \, dy + z^2 \, dz$, where C is the curve

$$\mathbf{r} = \mathbf{i} \cos t + \mathbf{j} \sin t + t\mathbf{k}, \ 0 \leq t \leq 2$$

Here $x'(t) = -\sin t$, $y'(t) = \cos t$ and $z'(t) = 1$, and the integral becomes

$$\int_0^2 (-\sin^2 t - \cos^2 t + t^2) \, dt = \int_0^2 (t^2 - 1) \, dt$$

$$= \frac{2}{3}$$

EXERCISES 7.2.1

1 Evaluate the line integral

$$\int_C xe^y \, dx + x^2 y \, dy$$

where C is the curve $\mathbf{r} = 3t\mathbf{i} + t^2\mathbf{j}, \ 0 \leq t \leq 1$.

2 Find $\int_C x \, dx - y \, dy + xy \, dz$, where C is the curve in Example 7.2.3.

7.3 PROPERTIES OF LINE INTEGRALS

If C is the curve $y = f(x)$, $x_1 \leq x \leq x_2$, shown in Figure 7.1(a), we have

$$\int_C \mathbf{F} \cdot d\mathbf{r} = \int_{x_1}^{x_2} (P(x, f(x)) + Q(x, f(x))f'(x)) \, dx$$

The properties of the definite integral (see Section 4.5) on the right-hand side of this equation lead to corresponding properties of the line integral. For example, we found that

$$\int_a^b f(x) \, dx = -\int_b^a f(x) \, dx$$

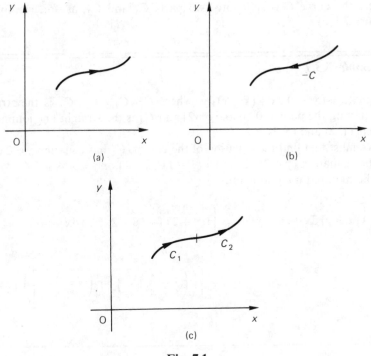

Fig. 7.1

We must therefore have

$$\int_{x_1}^{x_2} P(x)f(x))\,dx = -\int_{x_2}^{x_1} P(x, f(x))\,dx,$$

$$\int_{x_1}^{x_2} Q(x, f(x))f'(x)\,dx = -\int_{x_2}^{x_1} Q(x, f(x))f'(x)\,dx$$

and thus

$$\int_C \mathbf{F}\cdot d\mathbf{r} = -\int_{-C} \mathbf{F}\cdot d\mathbf{r}$$

where $-C$ denotes the curve C in reverse order (see Figure 7.1b). This result means that we must always specify the direction along which we integrate. Similarly, we find

$$\int_C \mathbf{F}\cdot d\mathbf{r} = \int_{C_1} \mathbf{F}\cdot d\mathbf{r} + \int_{C_2} \mathbf{F}\cdot d\mathbf{r}$$

207

where the curve C is split into two parts C_1 and C_2, as, for example, in Figure 7.1(c).

Example 7.3.1

Evaluate $\int_C (x + y)\,dx + (x - y)\,dy$, where $C = C_1 + C_2$, C_1 is the straight line joining the points $(0, 0)$ and $(1, 2)$ and C_2 is the straight line joining the points $(1, 2)$ and $(5, 3)$.

We must first find the equation of the curves. C_1 has equation $y = 2x$ and C_2 has equation $(y - 2)/(3 - 2) = (x - 1)/(5 - 1)$ or $y = x/4 + 7/4$.

The required integral is thus

$$\int_C (x + y)\,dx + (x - y)\,dy = \int_0^1 ((x + 2x) + (x - 2x)\cdot 2)\,dx$$

$$+ \int_1^5 \left(\left(x + \frac{x}{4} + \frac{7}{4}\right) + \frac{1}{4}\left(x - \frac{x}{4} - \frac{7}{4}\right)\right)dx$$

$$= 23$$

If the curve C is such that two distinct values of y are obtained for some value (or values) of x, as, for example, in Figure 7.2, then C must be split into C_1 and C_2, say, in such a way that in either part each value of x corresponds to just one value of y. In Figure 7.2, C is actually a closed curve, but the splitting of C into C_1 and C_2 enables us to evaluate a line integral along the whole of C. (Alternatively, if C can be specified parametrically, then the integral might be easier to evaluate as in Examples 7.2.2 and 7.2.3.)

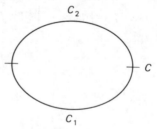

Fig. 7.2

Example 7.3.2

Evaluate $\int_C xy\,dx + ye^x\,dy$, where C is the perimeter of the rectangle shown in Figure 7.3.

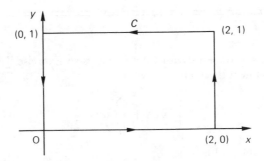

Fig. 7.3

Along $y = 0$ the integrand and, hence, the integral is zero. Along $y = 1$, y does not change, so $dy = 0$ and the second term in the integral contributes nothing. Similarly, along $x = 0$ and $x = 2$, $dx = 0$ so the first term does not contribute, while the first term in the integrand is zero along $x = 0$. We thus have altogether

$$\int_C xy\,dx + ye^x\,dy = \int_0^1 ye^2\,dy + \int_2^0 x\,dx + \int_1^0 y\,dy$$

$$= \tfrac{1}{2}e^2 - \tfrac{5}{2}$$

EXERCISES 7.3.1

1 Evaluate $\int_C (x + y)\,dx + (x - y)\,dy$ when C is
  (i) the triangle whose sides are $y = 0$, $x = 2$ and $y = \tfrac{1}{2}x$;
  (ii) the arc of the parabola $y = \tfrac{1}{4}x^2$ from $(0, 0)$ to $(2, 1)$.
2 Evaluate $\int_C y\,dx + x\,dy$ in the cases where
  (i) C is the line C_1 given by $y = x$, $0 \le x \le 1$;
  (ii) C is the curve C_2 given by $y = x^2$, $0 \le x \le 1$.

7.4 CONSERVATIVE FIELDS

If we move an object from A to B subject to the gravitational force field, we should expect to do the same amount of work whatever the path we use to get from A to B. However, it is not true that all line integrals are independent of path, as the following example shows.

Example 7.4.1

Evaluate $\int_C (x + y)\, dx + y\, dy$ for the two cases in Exercise 7.3.1.2.
   For (i), the integral equals

$$\int_0^1 2x\, dx + \int_0^1 y\, dy = \frac{3}{2}$$

For (ii), the integral is

$$\int_0^1 (x + x^2)\, dx + \int_0^1 y\, dy = \frac{4}{3}$$

On the other hand, Exercise 7.3.1.2 shows that the value of $\int_C y\, dy + x\, dy$ is the same when C is either C_1 or C_2. In fact, its value is the same whatever the curve C. The problem is to recognise what sort of vector fields possess this property; we now define a type of field which turns out to have just this property.

Definition

A vector field is called *conservative* if it is the gradient of a scalar field. If a vector field $\mathbf{V}$ is conservative, then a function f is a *potential* of $\mathbf{V}$ if $\nabla f = \mathbf{V}$.

Since the gradient of a constant is zero, the potential is only determined up to a constant, just like an indefinite integral. Gravity, static electricity and magnetic force fields are conservative. The negative of a potential function for fields of this kind is called a potential energy function, for obvious physical reasons. For example, the function f given by $f(x, y, z) = Gm/d$ is a potential function by Exercise 7.1.1.

We now show that, when $\mathbf{V}$ is a conservative vector field, the value of the line integral $\int_C \mathbf{V} \cdot d\mathbf{r}$ depends only on the end-points of C.

Let $\mathbf{V}(x, y) = P(x, y)\mathbf{i} + Q(x, y)\mathbf{j}$ be a conservative field over a region U of the plane and let f be a potential of $\mathbf{V}$. Let C be a piecewise smooth curve $y = g(x)$ inside U, which starts at the point $A = (x_1, g(x_1))$ and terminates at the point $B = (x_2, g(x_2))$. Then, since $\nabla f = \mathbf{V}$, we have $f_x = P$ and $f_y = Q$ and the line integral is

$$\int_C \mathbf{V} \cdot d\mathbf{t} = \int_C P(x, y)\, dx + Q(x, y)\, dy$$

$$= \int_{x_1}^{x_2} (f_x(x, g(x)) + f_y(x, g(x))g'(x))\, dx$$

But the integrand is $(d/dx)(f(x, g(x)))$, as we can see by the application of a chain rule. Thus,

$$\int_C \mathbf{V} \cdot d\mathbf{r} = \int_{x_1}^{x_2} (d/dx)(f(x, g(x))) \, dx$$

$$= f(x_2, g(x_2)) - f(x_1, g(x_1))$$

and the integral depends only on the values of f at the end-points of C, and not on the curve connecting them.

Note that if C is a closed curve, then $x_1 = x_2$ and the line integral, which is usually written as $\oint_C \mathbf{V} \cdot d\mathbf{r}$, has the value zero.

We can use the path independence of line integrals in conservative fields to choose a very simple curve along which to evaluate them. For example, if $A = (x_1, y_1)$ and $B = (x_2, y_2)$ are as above, we can choose C to consist of C_1, the line $y = y_1$ from $x = x_1$ to $x = x_2$, and C_2, the line $x = x_2$ from $y = y_1$ to $y = y_2$, as shown in Figure 7.4.

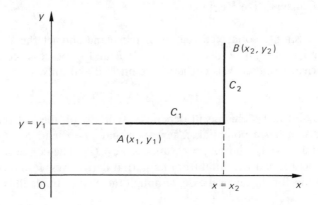

Fig. 7.4

We now have

$$\int_C P(x, y) \, dx + Q(x, y) \, dy = \left(\int_{C_1 + C_2} \right) P(x, y) \, dx + Q(x, y) \, dy$$

$$= \int_{x_1}^{x_2} P(x, y_1) \, dx + \int_{y_1}^{y_2} Q(x_2, y) \, dy$$

since dy is zero on $y = y_1$ and dx is zero on $x = x_2$.

It is apparent that, to make use of this property of a conservative field, we must be able to recognise one. We shall give a suitable test, which requires the region of the plane containing the curve C along which we evaluate the line integral to be *simply connected*. Roughly speaking, this means that the region has no 'holes' in it. A plane region is simply connected if every simple closed curve in the region can be continuously shrunk to a point. It is clear that the region shaded in Figure 7.5 is simply connected, while that in Figure 7.6 is not, since the curve C cannot shrink to a point because of the hole.

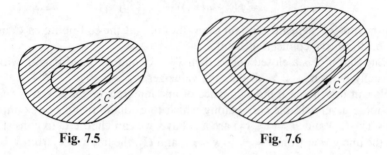

Fig. 7.5   Fig. 7.6

Test for Conservative Regions

Let U be a simply connected region of the plane and let the vector field $V = P(x, y)\mathbf{i} + Q(x, y)\mathbf{j}$ be defined on U. If P and Q possess second-order partial derivatives, then V is conservative on U if and only if

$$P_y(x, y) = Q_x(x, y)$$

for every point in U. This result depends essentially on the fact that if V is conservative, it has a potential, f, say, such that $f_x = P$ and $f_y = Q$, so that $P_y = f_{xy}$ and $Q_x = f_{yx}$; but $f_{xy} = f_{yx}$ and so $P_y = Q_x$. The converse requires the construction of a potential from its partial derivatives, P and Q, but we shall omit this for the general case, although the next example illustrates the idea.

Examples 7.4.2

1. Show that the vector field $V = 3x^2 y^2 \mathbf{i} + (y^3 + 2x^3 y)\mathbf{j}$ is conservative and find a potential function.

$$\frac{\partial}{\partial y}(3x^2 y^2) = 6x^2 y = \frac{\partial}{\partial x}(y^3 + 2x^3 y)$$

so V is conservative.

Let f be a potential function; then $f_x(x, y) = 3x^2 y^2$. Thus,

$$f(x, y) = \int 3x^2 y^2 \, dx = x^3 y^2 + g(y)$$

where we have treated y as a constant in integrating with respect to x, and added in an arbitrary function g of y, rather than a constant, since partial differentiation with respect to x will remove any such function. Now differentiate f partially with respect to y, to obtain

$$f_y(x, y) = 2x^3 y + g'(y)$$

But, from the expression for $\mathbf{V}$, this must equal $y^3 + 2x^3 y$, so we must have $g'(y) = y^3$ and, hence, $g(y) = \frac{1}{4}y^4 + \text{constant}$. The complete potential is thus $f(x, y) = x^3 y^2 + \frac{1}{4}y^4$, from which the constant is omitted, since we were only asked for a potential. The reader should check that $f_x = P$ and $f_y = Q$.

2. Let C be the curve $y^2 = x$, $0 \leq x \leq 1$. Evaluate the integral

$$\int_C (xy^2 + x^2)\, dx + (x^2 y + y^2)\, dy$$

Rather than solve the problem directly, we shall check that the field is conservative, then replace C by a simpler curve. Since

$$\frac{\partial}{\partial x}(x^2 y + y^2) = 2xy = \frac{\partial}{\partial y}(xy^2 + x^2)$$

the field is conservative. Now replace C, shown in Figure 7.7, by $C_1 + C_2$, where C_1 is the line $y = 0$ from $x = 0$ to $x = 1$ and C_2 is the line $x = 1$

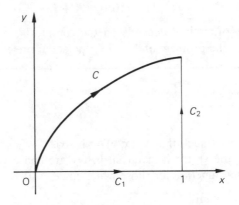

Fig. 7.7

213

from $y = 0$ to $y = 1$. The integral becomes

$$\int_{C_1} (xy^2 + x^2)\, dx + \int_{C_2} (x^2 y + y^2)\, dy$$

$$= \int_0^1 x^2\, dx + \int_0^1 (y + y^2)\, dy$$

$$= \frac{1}{3} + \frac{1}{2} + \frac{1}{3}$$

$$= \frac{7}{6}$$

Alternatively, we could find a potential f by letting

$$f_x(x, y) = xy^2 + x^2$$

and integrating this 'partially' with respect to x, to obtain

$$f(x, y) = \tfrac{1}{2}x^2 y^2 + \tfrac{1}{3}x^3 + g(y)$$

where g is an arbitrary function of y. Differentiating this partially with respect to y, we find

$$f_y(x, y) = x^2 y + g'(y)$$

If f is to be a potential, this must equal the second part of the integrand, $x^2 y + y^2$, and so we must have $g'(y) = y^2$ and, hence, $g(y) = \tfrac{1}{3}y^3 + c$. Leaving out the c, as before, we therefore obtain

$$f(x, y) = \tfrac{1}{2}x^2 y^2 + \tfrac{1}{3}(x^3 + y^3)$$

The required integral is now

$$f(1, 1) - f(0, 0) = \tfrac{1}{2} + \tfrac{2}{3} = \tfrac{7}{6}$$

The choice of method depends on the complexity of the curve C, on the context of the problem and also on personal preference.

EXERCISES 7.4.1

1 Find a function f such that $\nabla f(x, y) = y\mathbf{i} + x\mathbf{j}$.
2 Test whether the vector fields given below are conservative or not, and find a potential function for those which are.

   (i) $2xy\mathbf{i} + (x^2 + 1)\mathbf{j}$;
   (ii) $xe^y\mathbf{i} + ye^x\mathbf{j}$;

(iii) $(x \ln x + x^2 - y)\mathbf{i} - x\mathbf{j}$;

(iv) $(\sinh x \cosh y + 1)\mathbf{i} + (\cosh x \sinh y + x)\mathbf{j}$.

3 Evaluate the following line integrals:

(i) $\displaystyle\int_C (x^2 + y)\,dx + (y^2 - x)\,dy$

where C is the curve $\mathbf{r}(t) = (t^2, t^3)$, $0 \le t \le 1$;

(ii) $\displaystyle\int_C (e^x \cos y + e^y \cos x)\,dx + (e^y \sin x - e^x \sin y)\,dy$

where C is the curve $\mathbf{r}(t) = ((\pi/4)\sin t, (\pi/2)\sin t + \pi/2)$, $0 \le t \le \pi/2$;

(iii) $\displaystyle\oint_C (x^2 + y)\,dx + (x^2 + y^2)\,dy$

where C is the ellipse $x^2/4 + y^2/9 = 1$;

(iv) $\displaystyle\int_C (\cos y + y \cos x)\,dx + (y - x \sin y + \sin x)\,dy$

where C is the curve $\mathbf{r}(t) = (\cosh t, e^t)$, $0 \le t \le \ln 2$.

7.5 DOUBLE INTEGRALS

Recall that in Section 3.1 we defined the definite integral of a function f as the area under the graph of f; this was computed as the limit of a sum of areas of rectangles, whose width became vanishingly small.

We now carry out a similar construction for the double integral of a function f of two variables; in this case, we start with a region R of area A of the xy plane and construct a cylinder with R as its base, and with its axis parallel to the z axis, as shown in Figure 7.8. Then the integral of f over the region R is the volume of the cylinder between the xy plane and the surface $z = f(x, y)$. In order to compute this volume, we divide R into a number of subregions, the ith of which, δR_i, has area δA_i, and construct cylinders on each of these as a base with axes parallel to the z axis. If (x_i, y_i) is a point inside the subregion δR_i, then the volume between this and the surface $z = f(x, y)$ is approximately $f(x_i, y_i)\delta A_i$. Summing over all these subregions and proceeding to the limit as $\delta A_i \to 0$, which requires the number

215

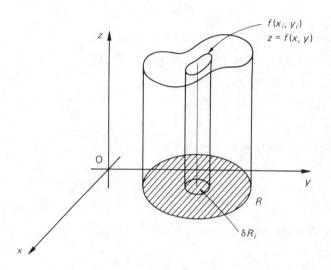

Fig. 7.8

of subregions to increase infinitely to maintain the equality

$$\sum_i \delta A_i = A$$

we obtain the double integral of the function f over the region R as

$$\iint_R f \, dA = \lim_{\max \delta A_i \to 0} \sum_i f(x_i, y_i) \delta A_i$$

We assume that the errors of approximating the cylinder over R by cylinders with flat tops disappear in the limit. We say that f is integrable if the limit exists; it can be shown to exist for suitably smooth functions. We now turn to a method of evaluating such integrals.

Repeated Integrals

We reduce the problem of evaluating a double integral to that of evaluating two single integrals (hence the term 'repeated integral'). We start by choosing the subregion of R, δR_{ij}, to be the rectangle whose sides are $x = x_i$, $x = x_i + \delta x_i$, $y = y_j$ and $y = y_j + \delta y_j$, where the subscripts i and j vary in such a way that the whole of R is covered. Figure 7.9 illustrates this. The area of δR_{ij} is clearly $\delta x_i \delta y_j$, so that we may express the double integral as

$$\iint_R f \, dA = \lim_{\max \delta x_i, \delta y_j \to 0} \sum_i \sum_j f(x_i, y_j) \delta x_i \delta y_j$$

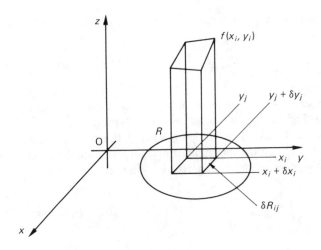

Fig. 7.9

where the limit is such that the whole of R is included, and must be taken with some care, to allow for the errors in approximating a non-rectangular region by rectangular subregions. When this limit exists, we write the double integral of f over R as

$$\iint_R f(x, y)\,dx\,dy$$

Fortunately, the limiting process works perfectly well for most smooth functions we shall meet, and we need not worry about the details; however, it is useful to realise that many of the properties of integrals derive from the properties of the corresponding sums whence they came.

Suppose we evaluate the sum over j first; we can show this by writing it as

$$\sum_i \left(\sum_j f(x_i, y_j)\delta y_j \right)\delta x_i \tag{1}$$

so that the quantity we must first evaluate for each value of i is

$$\sum_j f(x_i, y_j)\delta y_j \tag{2}$$

Now recall that, for the case of a definite integral, we have

$$\lim_{\delta y_j \to 0} \sum_j g(y_j)\delta y_j = \int_a^b g(y)\,dy \tag{3}$$

where a and b are the smallest and largest values taken by y. Since x_i is fixed

217

for each value of i, we may replace $g(y)$ by $f(x_i, y)$ in Equation (3), to obtain

$$\lim_{\delta y_j \to 0} \sum_j f(x_i, y_j)\delta y_j = \int_a^b f(x_i, y)\,dy$$

which we can use for the limit of Equation (2) with suitable values for a and b. In Figure 7.10 we have shown a typical line $x = x_i$; if the lower boundary of R has equation $y = g(x)$ and the upper boundary of R has equation $y = h(x)$, then the smallest and largest y values covered by the line $x = x_i$ are $g(x_i)$ and $h(x_i)$, respectively. We therefore write

$$\lim_{\delta y_j \to 0} \sum_j f(x_i, y_j)\delta y_j = \int_{g(x_i)}^{h(x_i)} f(x_i, y)\,dy = F(x_i),\ \text{say} \qquad (4)$$

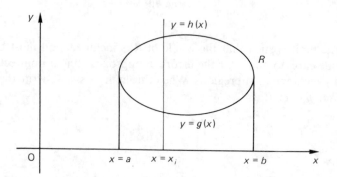

Fig. 7.10

Expression (1) now becomes $\sum_i F(x_i)\delta x_i$, and we now need to take the limit of this sum as $\delta x_i \to 0$. This is again a single sum and becomes in the limit the definite integral $\int_a^b F(x)\,dx$, where a and b must be the minimum and maximum values of x contained in R, in order to include the whole region. Substituting into this the expression for $F(x)$ from Equation (4), we finally obtain

$$\int_a^b \left(\int_{g(x)}^{h(x)} f(x, y)\,dy \right) dx \qquad (5)$$

This is called a *repeated integral*; we first evaluate the inner integral

$$\int_{g(x)}^{h(x)} f(x, y)\,dy = F(x)$$

218

noting that this is a 'partial' integration with respect to y, since, as the derivation shows, x is kept constant, then we evaluate the outer integral $\int_a^b F(x)\,dx$. We usually omit the brackets and write the integral as

$$\int_a^b \int_{g(x)}^{h(x)} f(x, y)\,dy\,dx \tag{6}$$

in which it is understood that the inner integral is evaluated first.

Let us now see what the effect is of summing in the opposite order, that is, summing over i first. The required sum is now

$$\sum_j \sum_i f(x_i, y_j)\delta x_i \delta y_j \tag{7}$$

and the inner sum to be evaluated first is

$$\sum_i f(x_i, y_j)\delta x_i$$

which in the limit becomes

$$\int_{k(y_j)}^{l(y_j)} f(x, y_j)\,dx = G(y_j), \text{ say} \tag{8}$$

where the limits arise from the equations of the left- and right-hand boundaries of R, $x = k(y)$ and $x = l(y)$, respectively (see Figure 7.11). The sum (7) now becomes $\sum_j G(y_j)\delta y_j$, and, going to the limit as $\delta y_j \to 0$, this becomes

$$\int_c^d G(y)\,dy = \int_c^d \left(\int_{k(y)}^{l(y)} f(x, y)\,dx \right) dy$$

$$= \int_c^d \int_{k(y)}^{l(y)} f(x, y)\,dx\,dy \tag{9}$$

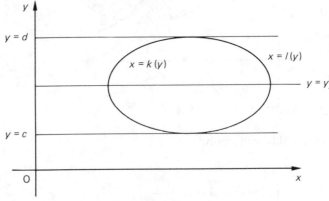

Fig. 7.11

Compared with the integral (5), we have 'reversed' the order of integration; since both integrals evaluate the same quantity, the volume of the cylinder under the surface $z = f(x, y)$, they must obviously have the same value.

Note that the limits of integration are completely different; we choose the order which makes the integration as easy as possible. When reversing the order of integration, a diagram is usually needed to find the changed limits.

Properties of Double Integrals

The following useful properties follow from the corresponding properties of finite sums:

(1) $\displaystyle\iint_R (f(x, y) + g(x, y))\,dx\,dy = \iint_R f(x, y)\,dx\,dy + \iint_R g(x, y)\,dx\,dy$

(2) $\displaystyle\iint_R kf(x, y)\,dx\,dy = k\iint_R f(x, y)\,dx\,dy$, where k is a constant

(3) $\displaystyle\iint_R f(x, y)\,dx\,dy = \iint_{R_1} f(x, y)\,dx\,dy + \iint_{R_2} f(x, y)\,dx\,dy$

where R_1 and R_2 are subregions of R, for example, as shown in Figure 7.12.

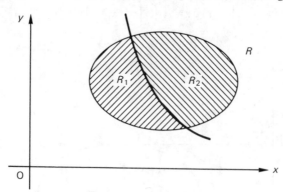

Fig. 7.12

We can write this result in briefer form as

$$\iint_R f(x, y)\,dx\,dy = \left(\iint_{R_1} + \iint_{R_2}\right) f(x, y)\,dx\,dy$$

220

Example 7.5.1

Evaluate $\iint_R (x^2 + y^2)\, \mathrm{d}A$, where R is the region of the xy plane bounded by $y = x^2$, $x = 2$ and $y = 1$.

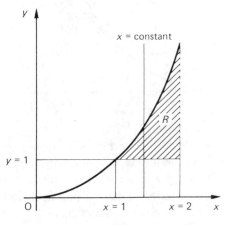

Fig. 7.13

We start by integrating with respect to y first. In order to find the limits, we must sketch the region R; this is done in Figure 7.13. Since we wish to integrate with respect to y first while keeping x constant, we draw a typical line $x = \text{constant}$. The limits for y are found from the intersection of this line with the lower and upper boundaries of the region of integration. These occur at $y = 1$ and $y = x^2$, so these are the y limits. To cover the whole of R, x must range from 1 to 2. Thus, the required integral is

$$\iint_R (x^2 + y^2)\, \mathrm{d}A = \int_{x=1}^{x=2} \left(\int_{y=1}^{y=x^2} (x^2 + y^2)\, \mathrm{d}y \right) \mathrm{d}x$$

where we have put in '$x =$' and '$y =$' to make quite clear what the limits refer to. Carrying out the inner integration with respect to y, we obtain

$$\iint_R (x^2 + y^2)\, \mathrm{d}A = \int_1^2 [x^2 y + \tfrac{1}{3}y^3]_1^{x^2}\, \mathrm{d}x$$

$$= \int_1^2 (\tfrac{1}{3}x^6 + x^4 - x^2 - \tfrac{1}{3})\, \mathrm{d}x$$

$$= [\tfrac{1}{21}x^7 + \tfrac{1}{5}x^5 - \tfrac{1}{3}x^3 - \tfrac{1}{3}x]_1^2$$

$$= \tfrac{1006}{105}$$

221

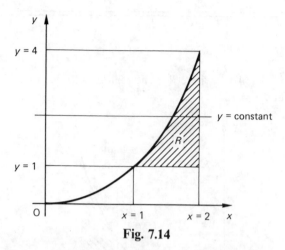

Fig. 7.14

Alternatively, if we decide to integrate with respect to x first, then the limits can be found from the intersection of a typical line $y = $ constant with the boundary of R. Such a line is shown in Figure 7.14; the required limits are clearly $x = \sqrt{y}$ and $x = 2$. Note that we had to solve for x in terms of y to obtain the appropriate x values. We now have

$$\iint_R (x^2 + y^2)\, dx\, dy = \int_{y=1}^{y=4} \left(\int_{x=\sqrt{y}}^{x=2} (x^2 + y^2)\, dx \right) dy$$

$$= \int_1^4 [\tfrac{1}{3}x^3 + y^2 x] \frac{2}{\sqrt{y}}\, dy$$

$$= \int_1^4 (\tfrac{8}{3} + 2y^2 - \tfrac{1}{3}y^{\frac{3}{2}} - y^{\frac{5}{2}})\, dy$$

$$= [\tfrac{8}{3} + \tfrac{2}{3}y^3 - \tfrac{2}{15}y^{\frac{5}{2}} - \tfrac{2}{7}y^{\frac{7}{2}}]_1^4$$

$$= \tfrac{1006}{105}$$

We now give an example where the region of integration has to be divided up to obtain suitable limits of integration.

Example 7.5.2

Evaluate $\iint_R (x - y)^2\, dA$, where R is the region defined by $0 \le y \le 1$ and $y \le x \le y + 2$.

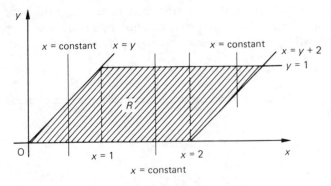

Fig. 7.15

The boundary lines of R are given by $y = 0$, $y = 1$, $x = y$ and $x = y + 2$, so that R is the interior of the parallelogram shown in Figure 7.15. Integrating with respect to y first, we see that there is not a completely typical line $x = $ constant, because the nature of its end-points changes as x changes. We therefore divide R into three subregions by the dashed lines. Typical lines $x = $ constant within each of these subregions show that the required limits are:

for $0 \leq x \leq 1$, y has limits 0 and x;
for $1 \leq x \leq 2$, y has limits 0 and 1;
for $2 \leq x \leq 3$, y has limits $x - 2$ and 1.

We thus have

$$\iint_R (x - y)^2 \, dA = \left(\int_0^1 \int_0^y + \int_1^2 \int_0^1 + \int_2^3 \int_{x-2}^1 \right)(x - y)^2 \, dy \, dx$$

If, however, we integrate first with respect to x, only one typical line $y = $ constant is required for the whole region. This intersects the left and right boundary of R in $x = y$ and $x = y + 2$, so the integral becomes

$$\iint_R (x - y)^2 \, dA = \int_0^1 \int_y^{y+2} (x - y)^2 \, dx \, dy$$

$$= \int_0^1 \left[\tfrac{1}{3}(x - y)^3 \right]_y^{y+2} \, dy$$

$$= \tfrac{8}{3} \int_0^1 \, dy$$

$$= \tfrac{8}{3}$$

The student should check that the first method does give the same answer. This example clearly shows that consideration of the order of integration is well worth while, since the second order gives a much easier solution.

Example 7.5.3

Evaluate

$$\int_0^a \int_x^a \frac{x \, dy \, dx}{\sqrt{x^2 + y^2}}$$

In this example the order of integration is already determined by the way it is written. If we proceed to evaluate it, however, we obtain

$$\int_0^a \left[x \sinh^{-1} \frac{y}{x} \right]_x^a dx = \int_0^a x \left(\sinh^{-1} \frac{a}{x} - \sinh^{-1} 1 \right) dx$$

and we have a difficult integral!

Let us try to reverse the order of integration; we shall need a sketch (as shown in Figure 7.16) in order to carry this out. The sketch in Figure 7.16 has been constructed by noting that the x limits are 0 and a, so that the lines $x = 0$ and $x = a$ must bound the region of integration, R; a line $x = $ constant must cut the boundary of R in $y = x$ and $y = a$. These equations therefore complete the specification of the boundary lines. In Figure 7.17 the typical line $y = $ constant cuts the boundary of R in $x = 0$ and $x = y$. The

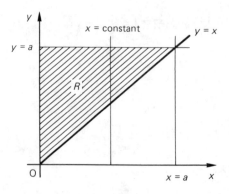

Fig. 7.16

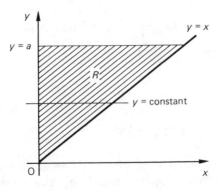

Fig. 7.17

integral thus becomes

$$\int_0^a \int_0^y \frac{x\,dx\,dy}{\sqrt{x^2+y^2}} = \int_0^a [\sqrt{x^2+y^2}]_0^y\,dy$$

$$= \int_0^a y(\sqrt{2}-1)\,dy$$

$$= \frac{\sqrt{2}-1}{2}a^2$$

EXERCISES 7.5.1

1 Sketch the region R enclosed by $y \le x+1$, $y \ge x^2+1$. Express $\iint_R xy\,dA$ as a repeated integral and evaluate it. Reverse the order of integration and re-evaluate it.

2 Sketch the region R enclosed by $y = \frac{1}{2}x$, $y = \frac{1}{2}x+1$, $y = 2x$ and $y = 2x+1$. Give limits for the integration of

$$\iint_R (y^2 - x^2)\,dy\,dx$$

but do not evaluate it.

3 For each of the following double integrals (a) evaluate it, (b) sketch the region of integration, (c) reverse the order of integration and re-evaluate it;

(i) $\displaystyle\int_0^a \int_0^{a-x} dy\,dx;$ (ii) $\displaystyle\int_0^a \int_0^x (x^2 + y^2)\,dy\,dx;$

(iii) $\displaystyle\int_0^1 \int_x^{\sqrt{x}} xy^2 \, dy \, dx$; (iv) $\displaystyle\int_{-\frac{\pi}{2}}^{\frac{\pi}{2}} \int_0^{2a\cos\theta} r^2 \cos\theta \, dr \, d\theta$.

7.6 CHANGE OF VARIABLES

One of the most powerful ways of evaluating an ordinary integral is to substitute a new variable for the variable of integration, with the aim of obtaining an easier function to integrate. The same technique can be used for double integrals, but, as we might expect, it is rather more complicated.

When we considered the method of substitution for definite integrals in Section 3.5, we used the substitution $x = g(u)$ to obtain

$$\int_a^b f(x) \, dx = \int_c^d f(g(u))g'(u) \, du$$

where $c = g^{-1}(a)$ and $d = g^{-1}(b)$. We can regard $g'(u)$ as a 'scaling factor', which is needed to compensate for the change of 'length' of dx caused by the change of variable. The function g must have a unique inverse in order for us to obtain the new limits.

Returning to the case of a double integral, we shall need a 'scaling factor', but it must now refer to areas. We also need to change two variables (in general, although there might be cases when changing one alone will suffice). Let us put $x = g(u, v)$ and $y = h(u, v)$; then, in order to find limits for u and v, we must find the region S of the uv plane which corresponds to the region of integration R in the xy plane and this will entail finding u and v in terms of x and y. A condition for there to be a unique solution is the non-vanishing of the Jacobian

$$\frac{\partial(x, y)}{\partial(u, v)} = \begin{vmatrix} \dfrac{\partial x}{\partial u} & \dfrac{\partial x}{\partial v} \\ \dfrac{\partial y}{\partial u} & \dfrac{\partial y}{\partial v} \end{vmatrix}$$

We shall see that this also provides the scaling factor we need. Before considering integrals further, we look at two examples of the effect of a change of variable on a region of the xy plane. We shall refer to this as a 'transformation' of variables, and to S as the region into which R is 'mapped' by the transformation.

Examples 7.6.1

1. Find the region S of the $r\theta$ plane corresponding to the region R defined by $x, y \geq 0$, $a^2 \leq x^2 + y^2 \leq b^2$, $0 < a < b$, when $x = r \cos \theta$ and $y = r \sin \theta$, $r \geq 0$, $0 \leq \theta \leq 2\pi$. What happens when $a = 0$?

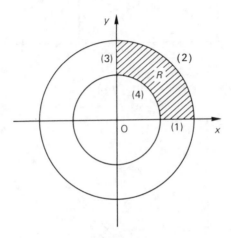

Fig. 7.18

The region R is the portion of the first quadrant between circles of radius a and b, centred at the origin, shown in Figure 7.18. We now derive what the various parts of the boundary of R, labelled as shown, correspond to (or map into) in the $r\theta$ plane:

(1) $y = 0$ maps into $\theta = 0$ (since $r = 0$ would mean that x and y are both zero);

(2) $x^2 + y^2 = b^2$ maps into $r = b$;

(3) $x = 0$ maps into $\theta = \pi/2$;

(4) $x^2 + y^2 = a^2$ maps into $r = a$.

Also, the point $x = y = (a+b)/2^{\frac{3}{2}}$ inside R corresponds to the point $r = (a+b)/2$, $\theta = \pi/4$, which is inside S. Thus, S is as shown in Figure 7.19. We note that S is a much easier region for which to specify limits of integration.

When $a = 0$, the curve (4) in the xy plane collapses to the point $(0, 0)$, which maps into the line $r = 0$ in the $r\theta$ plane. Such a point is called a *singular point* of the transformation.

2. Let R be defined by $y \geq 0$, $x + y \leq 6$ and $x - y \geq 2$. Find the region S into which R is mapped by the transformation $x = u + v$, $y = u - v$.

The boundary lines of R are shown numbered in Figure 7.20. They are

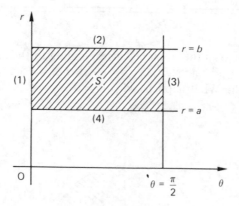

Fig. 7.19

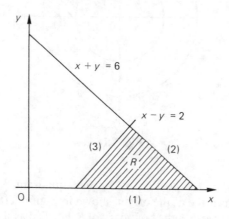

Fig. 7.20

transformed as follows:

(1) $y = 0$ maps into $u = v$;
(2) $x + y = 6$ maps into $u = 3$;
(3) $x - y = 2$ maps into $v = 1$;
the point (4,1)   maps into $(\frac{5}{2}, \frac{3}{2})$.

The results in (2) and (3) were obtained by substituting for x and y in terms of u and v.

The resulting region S is shown in Figure 7.21.

Suppose that we want to evaluate the double integral

$$\iint_R f(x, y) \, dx \, dy$$

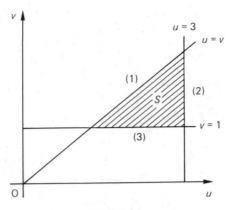

Fig. 7.21

Consider the transformation $x = g(u, v)$, $y = h(u, v)$, which maps the region R of the xy plane into the region S of the uv plane, as illustrated in Figures 7.22 and 7.23. We divide S into subregions, δS_{ij}, whose sides are (1) $u = u_i$, (2) $u = u_i + \delta u_i$, (3) $v = v_j$ and (4) $v = v_j + \delta v_j$, and i, j vary in such a way that the whole of S is covered. The region of the xy plane, δR_{ij}, which maps into δS_{ij} must have corresponding sides with parametric equations

(1) $x = g(u_i, v)$, $y = h(u_i, v)$;
(2) $x = g(u_i + \delta u_i, v)$, $y = h(u_i + \delta u_i, v)$;
(3) $x = g(u, v_j)$, $y = h(u, v_j)$;
(4) $x = g(u, v_j + \delta v_j)$, $y = h(u, v_j + \delta v_j)$.

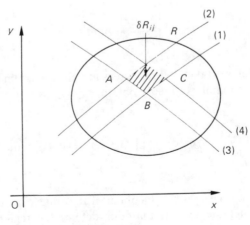

Fig. 7.22

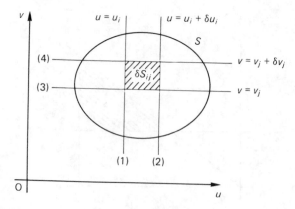

Fig. 7.23

We now attempt to evaluate the sum

$$\sum_i \sum_j f(g(u_i, v_j), h(u_i, v_j)) \delta A_{ij}$$

where δA_{ij} is the area of δR_{ij}. In the limit as $\delta A_{ij} \to 0$, this will give the required value of the double integral, since all we have done with the transformation is to use different subregions of R.

We now estimate the area δA_{ij}; the sides will not, in general, be straight, nor will opposite sides be parallel. However, we shall approximate δA_{ij} by the area of the parallelogram, three of whose vertices (those labelled A, B and C in Figure 7.22) coincide with those of δR_{ij}. We can use the result of Example 5.6.1.2 to compute the approximate area of this parallelogram, and after some fairly complicated manipulations, which we omit, this comes out as

$$\left| \frac{\partial(x, y)}{\partial(u, v)} \right| \delta u_i \delta v_j$$

where the modulus is taken because areas are always assumed to be positive. Proceeding to the limit, we finally obtain

$$\iint_R f(x, y)\, dx\, dy = \iint_S f(g(u, v), h(u, v)) |J|\, du\, dv$$

where $J = \partial(x, y)/\partial(u, v)$ is the Jacobian of the transformation. The reason we use S to integrate over is that this defines the region over which u and v vary.

Example 7.6.2

Evaluate $\iint_R \sqrt{x^2 + y^2}\, dx\, dy$, where R is the region $x, y \geq 0$, $a^2 \leq x^2 + y^2 \leq b^2$, $0 < a < b$, by transforming to polar coordinates.

This transformation was the subject of Example 7.6.1. The region R and the region S into which it is transformed are shown in Figures 7.18 and 7.19. With $x = r \cos \theta$ and $y = r \sin \theta$, the Jacobian is

$$\frac{\partial(x, y)}{\partial(r, \theta)} = \begin{vmatrix} \cos \theta & -r \sin \theta \\ \sin \theta & r \cos \theta \end{vmatrix} = r$$

and so

$$\iint_R \sqrt{x^2 + y^2}\, dx\, dy = \iint_S r^2\, dr\, d\theta$$

$$= \int_0^{\frac{\pi}{2}} \int_a^b r^2\, dr\, d\theta$$

$$= \frac{\pi}{6}(b^3 - a^3)$$

In this example, the transformation simplifies both the integrand and the region of integration.

It is often more convenient to define the inverse transformation, $u = l(x, y)$, $v = m(x, y)$. To evaluate the Jacobian, we can use the following result, which is a consequence of the appropriate chain rules:

$$\frac{\partial(x, y)}{\partial(u, v)} = 1 \bigg/ \frac{\partial(u, v)}{\partial(x, y)} = 1 \bigg/ \begin{vmatrix} \dfrac{\partial u}{\partial x} & \dfrac{\partial u}{\partial y} \\ \dfrac{\partial v}{\partial x} & \dfrac{\partial v}{\partial y} \end{vmatrix}$$

Example 7.6.3

Evaluate $\iint_R (x^2 + y^2)\, dx\, dy$, where R is the region of the xy plane bounded by $x^2 - y^2 = 1$, $x^2 - y^2 = 9$, $xy = 2$ and $xy = 4$.

The region R and the region S into which R is mapped by the transformation $u = x^2 - y^2$, $v = xy$ are shown in Figures 7.24 and 7.25. The Jacobian is

$$\frac{\partial(x, y)}{\partial(u, v)} = 1 \bigg/ \frac{\partial(u, v)}{\partial(x, y)} = 1 \bigg/ \begin{vmatrix} 2x & -2y \\ y & x \end{vmatrix} = \frac{1}{2(x^2 + y^2)}$$

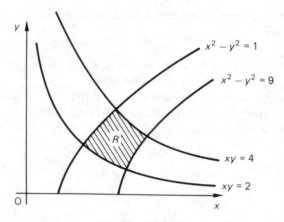

Fig. 7.24

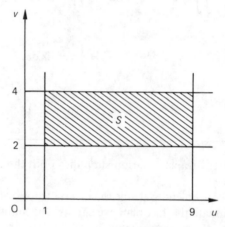

Fig. 7.25

The integral is now

$$\iint_S (x^2 + y^2)/(2(x^2 + y^2))\, du\, dv = \frac{1}{2}\int_2^4 \int_1^9 du\, dv$$

$$= 8$$

We end this section with a more practical example.

Example 7.6.4

Find the volume of the ellipsoid

$$\frac{x^2}{a^2} + \frac{y^2}{b^2} + \frac{z^2}{c^2} = 1$$

We find the volume by integrating the height over the region bounded by the cross-section of the ellipsoid in the plane $z = 0$ (think of the finite sum before going to the limit to obtain the integral. We add up the volume of all the tubes parallel to the z axis, which are contained in the ellipsoid, a typical one of which is shown in Figure 7.26). The required volume is

$$V = \iint_R 2z \, dx \, dy$$

where R is the interior of the ellipse $x^2/a^2 + y^2/b^2 = 1$. Note that the 2 is needed to include the lower half of the ellipsoid. Thus,

$$V = 2c \iint_R \sqrt{1 - \frac{x^2}{a^2} - \frac{y^2}{b^2}} \, dx \, dy$$

This integral can be evaluated by transforming to modified polar coordinates using $x = ar \cos \theta$, $y = br \sin \theta$, but we leave this as an exercise.

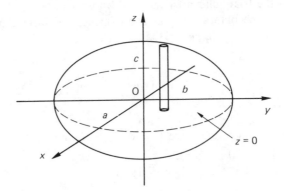

Fig. 7.26

EXERCISES 7.6.1

1 Sketch the region of integration of the integral

$$\int_0^1 \int_y^{2-y} ((x + y)/x^2) e^{x+y} \, dx \, dy$$

Transform the variables using $u = x + y$, $v = y/x$, and hence evaluate the integral.

2 Map the region R in Exercise 7.5.1.2 into the region S of the uv plane, using the transformation $u = 2x - y$, $v = x - 2y$. Transform the integral and hence evaluate it.

3 Evaluate the double integral

$$\iint_R (x^2 - 4xy - y^2)\,dx\,dy$$

where R is the region enclosed by $x \geq 0$, $y \geq 0$, $x^2 + y^2 \leq a^2$, by transforming to polar coordinates.

4 Show that the volume common to two right circular cylinders of radius a, whose axes are along the x and y axes, is given by

$$V = 16 \iint_R \sqrt{a^2 - x^2}\,dy\,dx$$

where R is the triangular area enclosed by $x = a$, $y = 0$ and $y = x$.

Show that the planes $z = \text{constant}$ intersect the surface of the above volume in squares, and, hence, that

$$V = 8 \int_0^a (a^2 - z^2)\,dz$$

Verify that both formulae give $V = \frac{16}{3}a^3$. (Note, no π!)

5 Show that $\int_0^\infty e^{-x^2}\,dx = \sqrt{\pi}/2$.
(*Hint*: Let $I = \int_0^\infty e^{-x^2}\,dx = \int_0^\infty e^{-y^2}\,dy$, so that

$$I^2 = \int_0^\infty e^{-x^2}\,dx \int_0^\infty e^{-y^2}\,dy = \int_0^\infty \int_0^\infty e^{-(x^2+y^2)}\,dx\,dy$$

Now transform to polar coordinates.)

7.7 GREEN'S THEOREM

We establish a connection between double integrals and line integrals, known as Green's theorem, which is very useful in physical applications.

Let R be a region enclosed by a smooth curve C, such as that shown in Figure 7.10. R lies between $x = a$ and $x = b$, and its lower and upper boundaries are defined by $y = g(x)$ and $y = h(x)$, respectively. Then

$$\iint_R P_y(x, y)\, dy\, dx = \int_a^b \int_{g(x)}^{h(x)} P_y(x, y)\, dy\, dx$$

$$= \int_a^b [P(x, y)]_{g(x)}^{h(x)}\, dx$$

$$= \int_a^b \{P(x, h(x)) - P(x, g(x))\}\, dx$$

$$= -\int_a^b P(x, g(x))\, dx - \int_b^a P(x, h(x))\, dx$$

$$= -\oint_C P(x, y)\, dx \qquad (1)$$

Similarly,

$$\iint_R Q_x(x, y)\, dx\, dy = \oint_C Q(x, y)\, dy \qquad (2)$$

Subtracting Equation (2) from Equation (1) gives

$$\iint_R \{P_y(x, y) - Q_x(x, y)\}\, dx\, dy = -\oint_C \{P(x, y)\, dx + Q(x, y)\, dy\} \qquad (3)$$

Special cases of Equations (1), (2) and (3) are obtained by taking $P(x, y) = y$ and $Q(x, y) = -x$. This gives

$$\iint_R P_y(x, y)\, dy\, dx = -\iint_R Q_x(x, y)\, dx\, dy = \iint_R dx\, dy = A$$

the area of R. Then

(1) gives $\quad A = -\oint_C y\, dx$;

(2) gives $\quad A = \oint_C x\, dy$; and

(3) gives $\quad 2A = \oint_C x\, dy - y\, dx$.

235

Example 7.7.1

Calculate the area enclosed by the ellipse $(x^2/a^2) + (y^2/b^2) = 1$.

We represent the ellipse (which we call C) parametrically by $x = a \cos t$, $y = b \sin t$, $0 \le t \le 2\pi$. Then the area is

$$A = \tfrac{1}{2} \oint_C x \, dy - y \, dx$$

$$= \frac{1}{2} \int_0^{2\pi} \{(a \cos t)(b \cos t) - (b \sin t)(-a \sin t)\} \, dt$$

$$= \frac{ab}{2} \int_0^{2\pi} (\sin^2 t + \cos^2 t) \, dt$$

$$= \frac{ab}{2} \int_0^{2\pi} dt$$

$$= ab\pi$$

Note that, if $a = b$, this gives the area of a circle of radius a.

EXERCISES 7.7.1

1 Find the area enclosed by the curve

$$r(t) = (\cos 2t \cos t, \cos 2t \sin t), \quad \frac{\pi}{4} \le t \le \frac{3\pi}{4}$$

(This is one leaf of the four-leaved curve in Exercise 5.9.1.4(v).)

2 Find the area of the triangle with vertices (a_1, a_2), (b_1, b_2) and (c_1, c_2). Compare your answer with that obtained using a vector product, as in Example 5.6.1.2.

Miscellaneous Exercises 7

For a thin wire in the shape of a curve C whose mass per unit length at a distance s along it is $m(s)$, the total mass is given by the line integral

$$M = \int_C m(s) \, ds$$

and the coordinates of its centre of mass are given by

$$\bar{x} = \frac{1}{M} \int_C xm(s)\,ds, \ \ \bar{y} = \frac{1}{M} \int_C ym(s)\,ds, \ \ \bar{z} = \frac{1}{M} \int_C zm(s)\,ds$$

Exercises 1 and 2 use these results.

1 A thin wire is in the shape of a semicircle $r(s) = (\cos s, \sin s)$, $0 \le s \le \pi$ and it has a mass per unit length $m(s) = 2 - \sin s$. Find the position of its centre of mass.

2 A thin wire of mass per unit length $m(t) = t$ is in the form of the curve

$$r(t) = \frac{t^2}{\sqrt{2}}i + \frac{t^2}{\sqrt{2}}j + \left(t - \frac{t^3}{3} \right)k, \ 0 \le t \le 1$$

Show that $ds/dt = 1 + t^2$, where s is the arclength of the curve, and use this to transform the integrals for M, $\bar{x}$, etc., into integrals with respect to t. Hence find the mass and the coordinates of the centre of mass of the wire.

For a thin flat plate occupying a region R of the xy plane, whose mass per unit area is $m(x, y)$ at the point (x, y), the total mass is given by

$$M = \iint_R m(x, y)\,dA$$

and the coordinates of the centre of mass by

$$\bar{x} = \frac{1}{M} \iint_R xm(x, y)\,dA \quad \text{and} \quad \bar{y} = \frac{1}{M} \iint_R ym(x, y)\,dA$$

When $m(x, y) = 1$ for all values of (x, y), the centre of mass is called the *centroid*. Exercises 3, 4 and 5 use these results.

3 Find the centroid of the semicircular region bounded by the x axis and the curve $y = \sqrt{1 - x^2}$.

4 Find the centroid of the region between the x axis and the curve $y = \sin x$, $0 \le x \le \pi$.

5 Find the centre of mass of a thin plate bounded by the curves $y = x^2$ and $y = 2x - x^2$ and whose mass per unit area at (x, y) is $m(x, y) = 1$.

6 A rectangular plate of unit mass per unit area occupies the region bounded by the lines $x = 0$, $x = 4$, $y = 0$, $y = 2$. The moment of inertia of the plate

about a line $y = a$ is given by the integral

$$I_a = \int_0^4 \int_0^2 (y - a)^2 \, dy \, dx$$

Find the value of a which minimises I_a.

7 A thin semicircular plate of radius a is immersed in a fluid with its plane vertical and its bounding diameter horizontal uppermost at a depth c. Show that the centre of pressure is at a depth

$$\frac{3\pi a^2 + 32ac + 12\pi c^2}{4(4a + 3\pi c)}$$

(*Hint*: Take the x axis along the bounding diameter and the y axis vertically downwards; the depth of the centre of pressure is then given by

$$\frac{\displaystyle\iint_R (c + y)^2 \, dA}{\displaystyle\iint_R (c + y) \, dA}$$

where the region of integration R is the surface of the plate.)

7.8 ANSWERS TO EXERCISES

Exercise 7.1.1

As in Example 7.1.2,

$$d_x = \frac{x}{d}, \; d_y = \frac{y}{d}, \text{ etc.}$$

Thus

$$f_x = -\frac{Gm}{d^2} d_x = -\frac{Gmx}{d^3}.$$

Similarly,

$$f_y = -\frac{Gmy}{d^3}, \; f_z = -\frac{Gmz}{d^3}.$$

Then

$$\nabla f = -\frac{Gm}{d^3}(x\mathbf{i} + y\mathbf{j} + z\mathbf{k}).$$

Exercises 7.2.1

1 Using the form (2), $x = 3t$, $y = t^2$, so $dx/dt = 3$, $dy/dt = 2t$, so the integral .
becomes $\int_0^1 (3te^{t^2} \cdot 3 + 9t^2 \cdot t^2 \cdot 2t) \, dt = [\frac{9}{2}e^{t^2} + 3t^6]_0^1 = \frac{9}{2}e - \frac{3}{2}$.
2 Following Example 7.2.3, the integral becomes $\int_0^{2\pi}(-\cos t \sin t - \sin t \cos t + \cos t \sin t) \, dt = -\frac{1}{2}\int_0^{2\pi} \sin 2t \, dt = \frac{1}{4}[\cos 2t]_0^{2\pi} = 0$.

Exercises 7.3.1

1 (i) $\displaystyle\int_C (x+y)\,dx + (x-y)\,dy = \int_0^2 x\,dx + \int_0^1 (2-y)\,dy + \int_2^0 \left(x + \frac{1}{2}x\right)dx$

$$+ \int_1^0 (2y - y)\,dy = 2 + \frac{3}{2} - 6 - \frac{1}{2} = -3.$$

(ii) $\displaystyle\int_0^2 \left(x + \frac{1}{4}x^2\right)dx + \int_0^1 (2\sqrt{y} - y)\,dy = \frac{7}{2}.$ $\left(\text{Or, parametrically, } x = 2t, \right.$

$\dfrac{dx}{dt} = 2$, $y = t^2 \cdot \dfrac{dy}{dt} = 2t$, so the integral becomes

$$\left. \int_0^1 \{(2t + t^2)2 + (2t - t^2)2t\}\,dt = \frac{7}{2}. \right)$$

2 (i) $\displaystyle\int_0^1 x\,dx + y\,dy = 1.$

(ii) $\displaystyle\int_0^1 x^2\,dx + \sqrt{y}\,dy = \frac{1}{3} + \frac{2}{3} = 1.$

Exercises 7.4.1

1 $f(x, y) = xy.$
2 (i) $(\partial/\partial y)2xy = 2x = (\partial/\partial x)(x^2 + 1)$, so conservative. Potential function,
$f(x, y) = \int 2xy\,dx = x^2 y + g(y)$, $f_y(x, y) = x^2 + g'(y)$, so $g'(y) = 1$,
$g(y) = y$. Thus $f(x, y) = (x^2 + 1)y$.
(ii) $(\partial/\partial y)xe^y = xe^y \neq ye^x = (\partial/\partial x)ye^x$, so not conservative.
(iii) $(\partial/\partial y)(x \ln x + x^2 - y) = -1 = (\partial/\partial x)(-x)$, so conservative. Potential
function, $f(x, y) = \int(x \ln x + x^2 - y)\,dx = \frac{1}{2}x^2 \ln x - \frac{1}{4}x^2 + \frac{1}{3}x^3 - yx + g(y)$, $f_y(x, y) = -x + g'(y) = -x$, so $g'(y) = 0$, $g(y) = a$
constant, which we may take as zero and $f(x, y) = \frac{1}{2}x^2 \ln x - \frac{1}{4}x^2 + \frac{1}{3}x^3 - xy$.

(iv) $(\partial/\partial y)(\sinh x \cosh y + 1) = \sinh x \sinh y \neq \sinh x \sinh y + 1 = (\partial/\partial x)(\cosh x \sinh y + x)$, so not conservative.

3 (i) Since $(\partial/\partial y)(x^2 + y) \neq (\partial/\partial x)(y^2 - x)$, the field is not conservative. Using the parametric form, the line integral becomes $\int_0^1 \{(t^4 + t^3)2t + (t^6 - t^2)3t^2\}\, dt = \frac{7}{15}$.

(ii) Here $(\partial/\partial y)(e^x \cos y + e^y \cos x) = -e^x \sin y + e^y \cos x = (\partial/\partial x)(e^y \sin x - e^x \sin y)$, so the field is conservative and we may integrate along any path from $(0, \frac{\pi}{2})$ to $(\frac{\pi}{4}, \pi)$. A potential is $f(x, y) = \int(e^x \cos y + e^y \cos x)\, dx = e^x \cos y + e^y \sin x + g(y)$, so $f_y(x, y) = e^y \cos x - e^x \sin y + g'(y)$, and $g'(y) = 0$, $g(y) = a$ constant, which we take as zero, giving $f(x, y) = e^x \cos y + e^y \sin x$. The value of the integral is $f(\frac{\pi}{4}, \pi) - f(0, \frac{\pi}{2}) = -e^{\frac{\pi}{4}} + e^{\pi}/\sqrt{2}$.

(iii) The field is not conservative here, so using the parametric form $\mathbf{r}(t) = (2 \cos t, 3 \sin t)$ for which $dx/dt = -2 \sin t$, $dy/dt = 3 \cos t$, the integral becomes $\int_0^{2\pi} \{(4 \cos^2 t + 3 \sin t)(-2 \sin t) + (4 \cos^2 t + 9 \sin^2 t)(3 \cos t)\}\, dt = [\frac{8}{3} \cos^3 t + \frac{3}{2} \sin 2t - 3t + 12 \sin t - 4 \sin^3 t + 9 \sin^3 t]_0^{2\pi} = -6\pi$.

(iv) $(\partial/\partial y)(\cos y + y \cos x) = -\sin y + \cos x = (\partial/\partial x)(y - x \sin y + \sin x)$, so conservative and we may integrate along any path from $(1, 1)$ to $(\cosh \ln 2, 2) = (\frac{5}{4}, 2)$ (see Example 1.9.2). We choose an easy path, as shown in Figure 7.4, and obtain for the integral $\int_1^{\frac{5}{4}}(\cos 1 + \cos x)\, dx + \int_1^2(y - \frac{5}{4} \sin y + \sin \frac{5}{4})\, dy = 2 \sin \frac{5}{4} + \frac{5}{4} \cos 2 - \sin 1 - \cos 1 + \frac{3}{2}$.

Exercises 7.5.1

1 The region R is shown below. $\int_0^1 \int_{x^2+1}^{x+1} xy\, dy\, dx = \frac{1}{2} \int_0^1 x[y^2]_{x^2+1}^{x+1}\, dx = \frac{1}{2} \int_0^1 (-x^3 + 2x^2 - x^5)\, dx = \frac{1}{8}$. In reverse order of integration, this is $\int_1^2 \int_{y-1}^{\sqrt{y-1}} xy\, dx\, dy = \frac{1}{2} \int_1^2 [x^2 y]_{y-1}^{\sqrt{y-1}}\, dy = \frac{1}{2} \int_1^2 (y^2 - y - y^3 + 2y^2 - y)\, dy = \frac{1}{8}$.

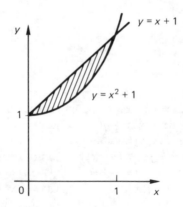

2 The region R is shown below. The integral is $\{\int_{-\frac{4}{3}}^0 \int_{\frac{1}{2}x}^{2x+1} + \int_0^2 \int_{2x}^{4x+1}\}(y^2 - x^2)\, dy\, dx$.

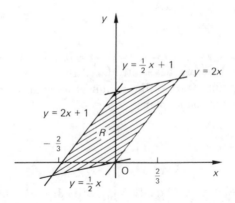

3 (i) (a) $\int_0^a (a-x)\,dx = \frac{1}{2}a^2$, (c) $\int_0^a \int_0^{a-y} dx\,dy = \int_0^a (a-y)\,dy = \frac{1}{2}a^2$.

(ii) (a) $\int_0^a [x^2 y + \frac{1}{3}y^3]_0^x \, dx = \frac{4}{3}\int_0^a x^3 \, dx = \frac{1}{3}a^4$, (c) $\int_0^a \int_y^a (x^2 + y^2)\,dx\,dy = \int_0^a [\frac{1}{3}x^3 + y^2 x]_y^a \, dy = \int_0^a (\frac{1}{3}a^3 - \frac{4}{3}y^3 - ay^2)\,dy = \frac{1}{3}a^4$.

(iii) (a) $\frac{1}{3}\int_0^1 [xy^3]_x^{\sqrt{x}} \, dx = \frac{1}{3}\int_0^1 (x^{\frac{5}{2}} - x^4)\,dx = \frac{1}{35}$, (c) $\int_0^1 \int_{y^2}^y xy^2 \, dx\,dy = \frac{1}{2}\int_0^1 [x^2 y^2]_{y^2}^y \, dy = \frac{1}{2}\int_0^1 (y^4 - y^6)\,dy = \frac{1}{35}$.

(iv) (a) $\frac{1}{3}\int_{-\frac{\pi}{2}}^{\frac{\pi}{2}} [r^3 \cos\theta]_0^{2a\cos\theta} \, d\theta = \frac{8a^3}{3}\int_{-\frac{\pi}{2}}^{\frac{\pi}{2}} \cos^4\theta \, d\theta = \frac{2a^3}{3}\int_{-\frac{\pi}{2}}^{\frac{\pi}{2}} (1 + \cos 2\theta)^2 \, d\theta = \pi a^3$, (c) $\int_0^{2a} \int_{-\alpha}^{\alpha} r^2 \cos\theta \, d\theta\,dr$, where $\alpha = \cos^{-1} x/2a$. This equals $\int_0^{2a} r^2 [\sin\theta]_{-\alpha}^{\alpha} \, dr = 2\int_0^{2a} r^2 \sqrt{1 - r^2/4a^2} \, dr$. This can be evaluated as πa^3 with the help of the substitution $r = 2a \sin u$.

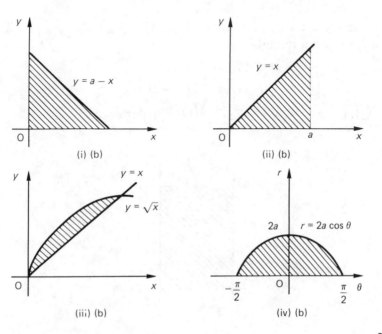

(i) (b)

(ii) (b)

(iii) (b)

(iv) (b)

Exercises 7.6.1

1   $y = x \to v = 1$, $x + y = 2 \to u = 2$, $y = 0 \to v = 0$, $x + y = 0 \to u = 0$. Note that for the last of these, we have introduced the line $x + y = 0$, which only contains one point, $(0,0)$, of the region of integration; it is transformed to the line $u = 0$ in the uv plane.

$$\frac{\partial(x, y)}{\partial(u, v)} = \frac{1}{\frac{\partial(u, v)}{\partial(x, y)}} = \begin{vmatrix} 1 & 1 \\ -\dfrac{y}{x^2} & \dfrac{1}{x} \end{vmatrix}^{-1} = \frac{x^2}{x + y}.$$

So $\int_0^1 \int_0^{2-y} ((x + y)/x^2) e^{x+y} \, dx \, dy = \int_0^1 \int_0^2 e^u \, du \, dv = e^2 - 1$.

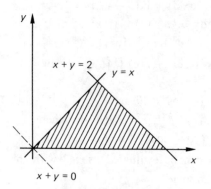

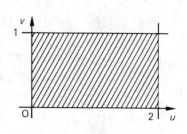

2   $y = 2x \to u = 0$, $y = 2x + 1 \to u = -1$, $y = \frac{1}{2}x \to v = 0$, $y = \frac{1}{2}x + 1 \to v = -2$, and S is just a rectangle.

$$\frac{\partial(u, v)}{\partial(x, y)} = \begin{vmatrix} 2 & -1 \\ 1 & -2 \end{vmatrix} = -3.$$

so $\iint_R (y^2 - x^2) \, dy \, dx = \int_{-2}^0 \int_{-1}^0 \frac{1}{3}(1/|-3|) \, du \, dv = \frac{2}{9}$.

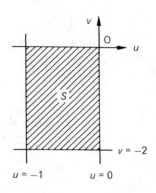

242

3 $x = r \cos \theta, \ y = r \sin \theta, \ \dfrac{\partial(x, y)}{\partial(r, \theta)} = \begin{vmatrix} \cos \theta & -r \sin \theta \\ \sin \theta & r \cos \theta \end{vmatrix} = r.$ Then

$$\iint\limits_{R} (x^2 - 4xy - y^2)\, dx\, dy = \int_0^{\frac{\pi}{2}} \int_0^{a} r^2(\cos 2\theta - 2 \sin 2\theta)r\, dr\, d\theta = -\frac{1}{2}a^4.$$

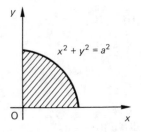
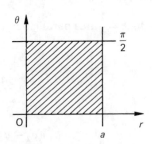

4 Because of symmetry, we need only integrate over the area shaded and for the upper half, and then multiply by 16. The height of the surface $x^2 + z^2 = a^2$ is $z = \sqrt{a^2 - x^2}$ and so $V = 16\iint_R \sqrt{a^2 - x^2}\, dy\, dx$, with R as given. Thus $V = 16\int_0^a \int_0^x \sqrt{a^2 - x^2}\, dy\, dx = 16\int_0^a x\sqrt{a^2 - x^2}\, dx = -\frac{16}{3}[(a^2 - x^2)^{\frac{3}{2}}]_0^a = \frac{16}{3}a^3.$ Planes $z = z_0$ intersect the surface in lines $x = \pm\sqrt{a^2 - z_0^2}, \ y = \pm\sqrt{a^2 - z_0^2},$ which form a square of area $4(a^2 - z_0^2)$. Integrating over positive z and allowing a factor of 2 for negative z gives the required result.

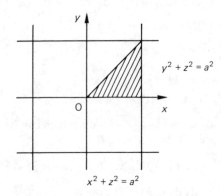

5 Using the results of Question 3, $I^2 = \int_0^\infty \int_0^\infty e^{-(x^2+y^2)}\, dx\, dy = \int_0^{\frac{\pi}{2}} \int_0^\infty e^{-r^2} r\, dr\, d\theta = -\frac{1}{2}\int_0^{\frac{\pi}{2}}[e^{-r^2}]_0^\infty\, d\theta = \frac{\pi}{4}.$

Exercises 7.7.1

1 $A = \frac{1}{2}\oint_c x\, dy - y\, dx = \frac{1}{2}\int_{\frac{\pi}{4}}^{\frac{3\pi}{4}}[\cos 2t \cos t\{-2 \sin 2t \sin t + \cos 2t \cos t\} - \cos 2t \sin t\{-2 \sin 2t \cos t - \cos 2t \sin t\}]\, dt = \frac{1}{2}\int_{\frac{\pi}{4}}^{\frac{3\pi}{4}} \cos^2 2t\, dt = \frac{\pi}{8}.$

243

2 Let $A = (a_1, a_2)$, $B = (b_1, b_2)$, $C = (c_1, c_2)$. Then the parametric equation of AB is $\mathbf{r}(t) = (a_1 + t(b_1 - a_1), a_2 + t(b_2 - a_2))$, $0 \le t \le 1$. Thus $\frac{1}{2}\int_{AB} x\,dy - y\,dx = \frac{1}{2}\int_0^1 [\{a_1 + t(b_1 - a_1)\}(b_2 - a_2) - \{a_2 + t(b_2 - a_2)\}(b_1 - a_1)]\,dt = \frac{1}{2}\int_0^1 (a_1 b_2 - a_2 b_1)\,dt = \frac{1}{2}(a_1 b_2 - a_2 b_1)$. Similarly, the contributions from BC and CA are $\frac{1}{2}(b_1 c_2 - b_2 c_1)$ and $\frac{1}{2}(c_1 a_2 - c_2 a_1)$, so the total area is $\frac{1}{2}(a_1 b_2 - a_2 b_1 + b_1 c_2 - b_2 c_1 + c_1 a_2 - c_2 a_1)$. From Example 5.6.1.2,

$$\text{area triangle} = \frac{1}{2}\text{area parallelogram} = \frac{1}{2}|\overrightarrow{AB} \times \overrightarrow{AC}|$$

$$= \frac{1}{2}\begin{vmatrix} \mathbf{i} & \mathbf{j} & \mathbf{k} \\ b_1 - a_1 & b_2 - a_2 & 0 \\ c_1 - a_1 & c_2 - a_2 & 0 \end{vmatrix}$$

$$= \frac{1}{2}|(b_1 - a_1)(c_2 - a_2) - (c_1 - a_1)(b_2 - a_2)|,$$

which expands to the answer above.

Miscellaneous Exercises 7

1 $M = \int_C (2 - \sin s)\,ds = [2s + \cos s]_0^\pi = 2\pi - 2$.

$\bar{x} = \frac{1}{2\pi - 2}\int_0^\pi \cos s(2 - \sin s)\,ds = \frac{1}{2\pi - 2}[2 \sin s - \frac{1}{2}\sin^2 s]_0^\pi = 0$.

$\bar{y} = \frac{1}{2\pi - 2}\int_0^\pi \sin s(2 - \sin s)\,ds = \frac{1}{2\pi - 2}[-2 \cos s - \frac{1}{2}s + \frac{1}{4}\sin 2s]_0^\pi = \frac{4 - \pi/2}{2\pi - 2}$.

2 $(ds/dt)^2 = (dx/dt)^2 + (dy/dt)^2 + (dz/dt)^2 = 2t^2 + 2t^2 + (1 - t^2)^2$

$= (1 + t^2)^2$, so $ds/dt = 1 + t^2$.

$M = \int_0^1 t(1 + t^2)\,dt = \frac{3}{4}$.

$\bar{x} = \bar{y} = \frac{4}{3}\int_0^1 (t^2/\sqrt{2})t(1 + t^2)\,dt = \frac{5}{9}\sqrt{2}$.

$\bar{z} = \frac{4}{3}\int_0^1 (t - t^3/3)t(1 + t^2)\,dt = \frac{176}{315}$.

3 $M = \int_{-1}^1 \int_0^{\sqrt{1-x^2}} dy\,dx = \int_{-1}^1 \sqrt{1 - x^2}\,dx$, which can be evaluated as $\frac{\pi}{2}$ with the help of the substitution $x = \sin \theta$.

$\bar{x} = \frac{2}{\pi}\int_{-1}^1 \int_0^{\sqrt{1-x^2}} x\,dy\,dx = \frac{2}{\pi}\int_{-1}^1 x\sqrt{1 - x^2}\,dx = \frac{2}{\pi}[-\frac{1}{3}(1 - x^2)^{3/2}]_{-1}^1 = 0$.

$\bar{y} = \frac{2}{\pi}\int_{-1}^1 \int_0^{\sqrt{1-x^2}} y\,dy\,dx = \frac{1}{\pi}\int_{-1}^1 (1 - x^2)\,dx = \frac{4}{3\pi}$.

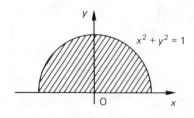

4 $M = \int_0^\pi \int_0^{\sin x} dy \, dx = \int_0^\pi \sin x \, dx = 2.$

$\bar{x} = \frac{1}{2}\int_0^\pi \int_0^{\sin x} x \, dy \, dx = \frac{1}{2}\int_0^\pi x \sin x \, dx = \frac{1}{2}[-x \cos x + \sin x]_0^\pi = \frac{\pi}{2}.$

$\bar{y} = \frac{1}{2}\int_0^\pi \int_0^{\sin x} y \, dy \, dx = \frac{1}{4}\int_0^\pi \sin^2 x \, dx = \frac{\pi}{8}.$

5 $M = \int_0^1 \int_{x^2}^{2x-x^2} dy \, dx = 2\int_0^1 (x - x^2) \, dx = \frac{1}{3}.$

$\bar{x} = 3\int_0^1 \int_{x^2}^{2x-x^2} x \, dy \, dx = 6\int_0^1 (x^2 - x^3) \, dx = \frac{1}{2}.$

$\bar{y} = 3\int_0^1 \int_{x^2}^{2x-x^2} y \, dy \, dx = \frac{3}{2}\int_0^1 (4x^2 - 4x^3) \, dx = \frac{1}{2}.$

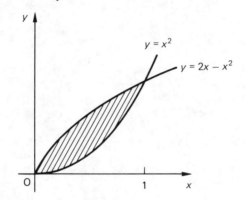

6 $I_a = \frac{1}{3}\int_0^4 [(y - a)^3]_0^2 \, dx = \frac{4}{3}\{(2 - a)^3 - a^3\}.$

For a minimum, $0 = dI_a/da = 4\{-(2 - a)^2 + a^2\} \Rightarrow a = 1.$

7 The depth of the centre of pressure is

$$\frac{\displaystyle\int_{-a}^{a} \int_0^{\sqrt{a^2 - x^2}} (c + y)^2 \, dy \, dx}{\displaystyle\int_{-a}^{a} \int_0^{\sqrt{a^2 - x^2}} (c + y) \, dy \, dx} = \frac{\dfrac{1}{3}\displaystyle\int_{-a}^{a} \{(c + \sqrt{a^2 - x^2})^3 - c^3\} \, dx}{\dfrac{1}{2}\displaystyle\int_{-a}^{a} \{(c + \sqrt{a^2 - x^2})^2 - c^2\} \, dx}$$

$$= \frac{2\displaystyle\int_{-a}^{a} \{(a^2 - x^2)^{3/2} + 3c(a^2 - x^2) + 3c^2(a^2 - x^2)^{1/2}\} \, dx}{3\displaystyle\int_{-a}^{a} \{a^2 - x^2 + 2c(a^2 - x^2)^{1/2}\} \, dx}$$

$$= \frac{2}{3}\left(\frac{\dfrac{3\pi a^4}{8} + 4ca^3 + \dfrac{3}{2}\pi c^2 a^2}{\dfrac{4}{3}a^3 + c\pi a^2}\right),$$

where the integrals involving the square roots were evaluated with the help of the substitution $x = a \sin \theta$. Cancellation produces the given answer.

8 COMPLEX NUMBERS

8.1 INTRODUCTION

Consider the four number sets, $\mathbb{N}$, $\mathbb{Z}$, $\mathbb{Q}$ and $\mathbb{R}$. Each one has algebraic shortcomings, illustrated by the following table:

Equation	Solution in:			
	$\mathbb{N}$	$\mathbb{Z}$	$\mathbb{Q}$	$\mathbb{R}$
$x + 1 = 0$	no	yes	yes	yes
$2x - 1 = 0$	no	no	yes	yes
$x^2 - 2 = 0$	no	no	no	yes
$x^2 + 1 = 0$	no	no	no	no

Each of the sets $\mathbb{N}$, $\mathbb{Z}$, $\mathbb{Q}$, and $\mathbb{R}$ includes the previous set and extends the class of equations for which solutions exist. The set $\mathbb{C}$ of complex numbers completes the extension to include solutions of the fourth equation, and indeed a much wider class of equations. This is demonstrated in a remarkable theorem, due to Gauss and known as the Fundamental Theorem of Algebra, which proves that every polynomial equation with coefficients in $\mathbb{C}$ has a solution in $\mathbb{C}$.

The significance of the complex numbers lies not only in the algebraic properties described above. If we think of the set of real numbers as one-dimensional, with the numbers corresponding to points on a line, then the set of complex numbers is two-dimensional with the numbers corresponding to points in a plane. This provides a mathematical tool for modelling physical problems involving two variables as problems of a single complex variable. This tool is both powerful in its applications and elegant in its purely mathematical nature. We start by investigating the algebraic properties of complex numbers.

8.2 THE ALGEBRA OF COMPLEX NUMBERS

Complex numbers first appeared in the sixteenth century in the work of the Italian mathematician Bombelli. They arose in connection with the solutions of equations, cubic as well as quadratic, and appeared in the form

$$x + iy \tag{1}$$

where x and y are real numbers and i is a 'number' whose square is -1. The existence and nature of this 'number' i gave rise to a certain amount of mystery and scepticism, some of which lingers on through the use of the term 'imaginary number'. We shall give a definition of complex number which overcomes the problem of existence. In practice we think of complex numbers in the above form and we shall see how our definition is consistent with this.

Definition

A *complex number* is an ordered pair (x, y) of real numbers, usually written as $x + iy$.

The use of the word 'ordered' in the definition is important; the ordered pairs (x, y) and (y, x) are distinct unless $x = y$.

The set of all complex numbers is denoted by $\mathbb{C}$ and it is usual to denote a typical member of the set by the single letter z. This is often more convenient in manipulations than the fuller form $x + iy$. Complex numbers can be represented as points in the Cartesian plane in an obvious way: $x + iy$ is represented by the point with Cartesian coordinates (x, y). For example, the number $2 - 3i$ (note that we write the second number, -3, in front of the i) is plotted as the point $(2, -3)$. We usually refer to the plane as the *Argand*

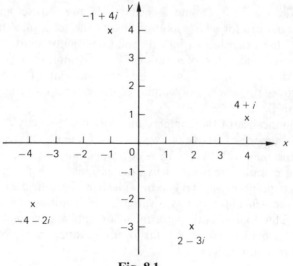

Fig. 8.1

diagram in this context. Figure 8.1 shows an example, with the point $2 - 3i$ and other points marked on it.

The complex number $0 + 1i$ is usually written as i. If x is a real number then we write it in the form of a complex number as $x + 0i$. Thus, in the Argand diagram, real numbers appear on the x axis, sometimes called the real axis for this reason.

Let $z = x + iy$. Then we call x and y the *real* and *imaginary* parts of z, and denote them by

$$x = \text{Re}(z) \quad \text{and} \quad y = \text{Im}(z)$$

Addition and Multiplication in $\mathbb{C}$

Let $z = x + iy$ and $z' = x' + iy'$ be complex numbers. Their sum and product are defined as follows:

$$z + z' = (x + x') + i(y + y')$$

$$zz' = (xx' - yy') + i(xy' + x'y)$$

We note that these definitions include the rules for adding and multiplying real numbers. For if y and y' are zero then z and z' are both real numbers and the rules give

$$z + z' = (x + x') + i(0 + 0) = x + x'$$

$$zz' = (xx' - 0) + i(0 + 0) = xx'$$

which are simply the sum and product of the two real numbers $z = x$ and $z' = x'$.

To find what happens when one of the two numbers is real and one complex, we let $y = 0$, so that z is real. Then

$$z + z' = (x + x') + i(0 + y') = (x + x') + iy'$$

$$zz' = (xx' - 0) + i(xy' + 0) = xx' + ixy'$$

Now we see what happens if the two numbers are both *imaginary*, that is, their real parts are zero. With x and x' both zero, we have

$$z + z' = (0 + 0) + i(y + y') = 0 + i(y + y')$$

and the sum is also imaginary; also

$$zz' = (0 - yy') + i(0 + 0) = -yy'$$

so the product of two imaginary numbers is *real*. In the particular case when $z = z' = i$, that is when $y = y' = 1$, we have

$$i^2 = zz' = -yy' = -1$$

This shows that the solution of $z^2 + 1 = 0$ in $\mathbb{C}$ is $z = \pm i$; thus, our definition of complex numbers is reconciled with the historical form (1).

It now becomes clear that the definitions of addition and multiplication are exactly what we should expect, since the quantity i can be manipulated like a real number. The sum of two complex numbers is found by collecting the terms in i to form the imaginary part, while the remaining terms form the real part. The product of two complex numbers comes from multiplying them term by term, replacing i^2 by -1 and collecting terms. This provides the easiest way of computing sums and products.

Example 8.2.1

Find $zz' - z''$ in the form $x + iy$ when $z = 1 - 2i$, $z' = 3 + 4i$ and $z'' = 2 + 5i$.

$$zz' - z'' = (1 - 2i)(3 + 4i) - (2 + 5i)$$
$$= 3 + 4i - 6i - 8i^2 - 2 - 5i$$
$$= 3 + 4i - 6i + 8 - 2 - 5i$$
$$= 9 - 7i$$

The student should check that the following laws hold for complex numbers:

$z + z' = z' + z$ (commutative law for addition)

$zz' = z'z$ (commutative law for multiplication)

$z + (z' + z'') = (z + z') + z''$ (associative law for addition)

$z(z'z'') = (zz')z''$ (associative law for multiplication)

$z(z' + z'') = zz' + zz''$ (distributive law for multiplication over addition)

For example, to prove the associative law for multiplication, with $z = x + iy$, etc.,

$$z(z'z'') = (x + iy)((x'x'' - y'y'') + i(x'y'' + x''y'))$$
$$= (x(x'x'' - y'y'') - y(x'y'' + x''y'))$$
$$\quad + i(x(x'y'' + x''y') + y(x'x'' - y'y''))$$
$$= ((xx' - yy')x'' - (xy' + x'y)y'')$$
$$\quad + i((xx' - yy')y'' + (xy' + x'y)x'')$$
$$= ((xx' - yy') + i(xy' + x'y))(x'' + iy'')$$
$$= (zz')z''$$

The real numbers 0 and 1 have special properties which they retain in the wider context of complex numbers. These are $z + 0 = z$ and $z \cdot 1 = z$ for all complex numbers z, and can easily be verified. Thus, 0 and 1 are the zero and unit of the complex numbers also.

Subtraction

Let $z = x + iy$ be a complex number. The complex number $(-x) + i(-y)$ is written $-z$. Subtracting z' from z means adding $-z'$ to z. Thus,

$$z - z' = z + (-z')$$
$$= (x - x') + i(y - y')$$

Let us interpret addition and subtraction in the Argand diagram. We can regard z and z' as the position vectors of the points X and X' whose coordinates are (x, y) and (x', y') respectively. The rule for the addition of complex numbers is identical to that for vectors; thus the parallelogram rule for vector addition (see Section 5.3) applies also to complex numbers, and is illustrated in Figure 8.2.

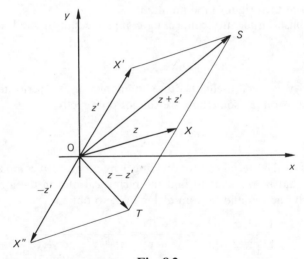

Fig. 8.2

To find $z - z'$, we complete the parallelogram with sides OX and OX'', where X'' has position vector $-z'$.

Conjugate and Modulus

Definitions

Let $z = x + iy$ be a complex number. Then the *complex conjugate* of z is $\bar{z} = x - iy$. The *modulus* of z is the non-negative square root of $x^2 + y^2$ and is written as $|z|$. Clearly $|z| > 0$.

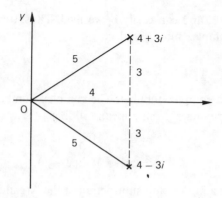

Fig. 8.3

For example, if $z = 4 + 3i$ then $\bar{z} = 4 - 3i$ and $|z| = 5$. The number $4 + 3i$ and its complex conjugate are shown in an Argand diagram in Figure 8.3.

It is obvious that $-z$ is the reflection of z in the x axis, while $|z|$ is the distance from the origin to the point z or $\bar{z}$.

We establish an identity connecting complex conjugate and modulus as follows:

$$z\bar{z} = (x + iy)(x - iy) = x^2 + y^2 = |z|^2$$

This identity is very useful because it enables us to perform algebraic manipulations on $|z|$, something we cannot do directly.

Division

We start with the special case of dividing 1 by a non-zero complex number $z = x + iy$, that is, we want to find the reciprocal of z in the form $x' + iy'$. We multiply the top and bottom of $1/z$ by $\bar{z}$, to obtain

$$\frac{1}{z} = \frac{1}{z} \cdot \frac{\bar{z}}{\bar{z}} = \frac{\bar{z}}{|z|^2} = \frac{x - iy}{x^2 + y^2} = \frac{x}{x^2 + y^2} - \frac{iy}{x^2 + y^2}$$

In a numerical example, we usually copy this procedure rather than remember the form of the reciprocal.

Example 8.2.2

Find the reciprocal of $2 - 3i$.

$$\frac{1}{2 - 3i} \cdot \frac{2 + 3i}{2 + 3i} = \frac{2 + 3i}{4 + 9} = \frac{2}{13} + \frac{3}{13}i$$

We tackle the problem of finding z'/z, $z \neq 0$, in a similar way,

$$\frac{z'}{z} = \frac{z'}{z} \cdot \frac{\bar{z}}{\bar{z}} = \frac{z'\bar{z}}{|z|^2}$$

Again, in a numerical example we just copy this procedure with z and z' replaced by their numerical values.

Example 8.2.3

Express the complex number $\dfrac{2+3i}{1+4i}$ in the form $x + iy$.

$$\frac{2+3i}{1+4i} = \frac{2+3i}{1+4i} \cdot \frac{1-4i}{1-4i} = \frac{14}{17} - \frac{5}{17}i$$

EXERCISES 8.2.1

1 Find in the form $x + iy$ the sum and product of the numbers $2 - 3i$ and $4 + 5i$. Show these numbers and their sum and product on an Argand diagram.

2 Simplify $i(2 - 9i) - (1 + i)(1 + 3i) - 5i$.

3 Find i^{-1} and $(1 + 4i)^{-1}$.

4 Express the following numbers in the form $x + iy$:

(i) $\dfrac{12 - 11i}{1 + 2i}$; (ii) $\dfrac{4 + 7i}{2 + 6i}$.

8.3 SOLUTION OF EQUATIONS

If the equation being solved is linear in z, that is, the maximum degree of z is 1, then it should be solved directly, just as though z were a real variable. For example, if a, b and c are complex numbers, $a \neq 0$, then the equation $az + b = c$ would have the solution $z = (c - b)/a$.

Quadratic Equations

We want to find the roots (that is, the solutions) of the quadratic equation

$$az^2 + bz + c = 0$$

where a, b and c are complex numbers, $a \neq 0$. We can do this as follows, starting by dividing through by a:

$$z^2 + \frac{b}{a}z + \frac{c}{a} = 0$$

'Completing the square', we obtain

$$\left(z + \frac{b}{2a}\right)^2 = \frac{b^2 - 4ac}{4a^2}$$

and taking square roots of both sides gives us $z + \dfrac{b}{2a} = \pm\dfrac{\sqrt{b^2 - 4ac}}{2a}$

Finally, rearranging, we obtain the familiar formula

$$z = \frac{-b \pm \sqrt{b^2 - 4ac}}{2a}$$

If a, b and c are real, we have real distinct roots, real coincident roots or a conjugate complex pair of roots depending on whether $b^2 - 4ac$ is positive, zero or negative. In the case where a, b and c are complex we have to find the square root of the complex number $b^2 - 4ac$, for which we shall develop a technique immediately. In Section 8.6, we shall see another method, which also works for the solution of $z^n = a$, where n is any natural number.

A useful tool in solving equations with complex solutions is found by using the fact that two complex numbers are equal if and only if their real and imaginary parts are each equal. This leads to the technique of 'equating real and imaginary parts'.

Example 8.3.1

Solve the equation $z^2 = 3 - 4i$.
Put $z = x + iy$ in the equation to obtain

$$(x^2 - y^2) + 2xyi = 3 - 4i$$

Equating the real parts on the left and right gives the equation

$$x^2 - y^2 = 3$$

and equating imaginary parts gives

$$2xy = -4$$

The second equation gives $y = -(2/x)$, and putting this into the first equation, we obtain

$$x^2 - \frac{4}{x^2} = 3$$

which rearranges to

$$x^4 - 3x^2 - 4 = 0$$

This is a quadratic equation in x^2 with roots $x^2 = 4$ or -1, but as x is real, we must have $x = \pm 2$. Since $y = -(2/x)$, it follows that $y = \mp 1$ and so $z = \pm(2 - i)$.

We can find the square root of any complex number in this way, and hence solve any quadratic equation.

EXERCISES 8.3.1

1 Find all solutions in $\mathbb{C}$ of the equations

(i) $z^2 = -8 + 6i$; (ii) $(1 + i)z + (2 - i) = i$;
(iii) $|z| - \bar{z} = 1 + 2i$; (iv) $\bar{z} = z^2$.

2 Show that the equation $z^2 = a = b + ic$, $c > 0$, has solution

$$z = \pm \left[\sqrt{\frac{|a| + b}{2}} + i \sqrt{\frac{|a| - b}{2}} \right]$$

3 Solve the equation $z^2 - (1 + 4i)z - (3 - i) = 0$.
4 Solve the equation $z^4 - 2z^2 + 4 = 0$. (*Hint*: this is a quadratic in z^2.)

8.4 EQUALITIES AND INEQUALITIES

The following is a list of useful properties, a few of which we shall verify and the rest of which we shall leave the verification of as an exercise (the first we have already seen, but is included for completeness):

$$z\bar{z} = |z|^2 \tag{1}$$

$$z = \text{Re}(z) + i\,\text{Im}(z) \tag{2}$$

$$\text{Re}(z + z') = \text{Re}(z) + \text{Re}(z') \tag{3}$$

$$\text{Im}(z + z') = \text{Im}(z) + \text{Im}(z') \tag{4}$$

$$\text{Re}(z) = (z + \bar{z})/2 \tag{5}$$

$$\text{Im}(z) = (z - \bar{z})/2i \tag{6}$$

$$|\text{Re}(z)| \leq |z| \tag{7}$$

$$|\text{Im}(z)| \leq |z| \tag{8}$$

$$z = \bar{z} \text{ if and only if } z \text{ is real} \tag{9}$$

$$z = -\bar{z} \text{ if and only if } z \text{ is imaginary} \tag{10}$$

$$\text{if } z \neq 0 \text{ then } |z| \neq 0 \text{ and } \bar{z} \neq 0 \tag{11}$$

$$|z| = |\bar{z}| = |-z| = |-\bar{z}| \tag{12}$$

$$\overline{z + z'} = \bar{z} + \bar{z}' \tag{13}$$

$$\overline{z - z'} = \bar{z} - \bar{z}' \tag{14}$$

$$\bar{\bar{z}} = z \tag{15}$$

$$\overline{zz'} = \bar{z}\bar{z}' \tag{16}$$

$$\overline{\left(\frac{z}{z'}\right)} = \frac{\bar{z}}{\bar{z}'} \quad (z' \neq 0) \tag{17}$$

$$|zz'| = |z||z'| \tag{18}$$

$$\left|\frac{z}{z'}\right| = \frac{|z|}{|z'|} \quad (z' \neq 0) \tag{19}$$

$$|z + z'| \leq |z| + |z'| \quad \text{(triangle inequality)} \tag{20}$$

Proof of (7) Let $z = x + iy$. Then

$$|z|^2 = x^2 + y^2 \geq x^2 = (\text{Re}(z))^2$$

Proof of (16) Let $z = x + iy$ and $z' = x' + iy'$. Then

$$\overline{zz'} = (xx' - yy') - i(xy' + x'y)$$

$$= (x - iy)(x' - iy')$$

$$= \bar{z}\bar{z}'$$

Proof of (17) $z'\dfrac{z}{z'} = z$, so $\bar{z}'\overline{\left(\dfrac{z}{z'}\right)} = \bar{z}$. Since $z' \neq 0$ then $\bar{z}' \neq 0$ so

$$\overline{\left(\frac{z}{z'}\right)} = \frac{\bar{z}}{\bar{z}'}$$

Proof of (18)

$$|zz'|^2 = zz'\overline{zz'}$$

$$= z\bar{z}z'\bar{z}'$$

$$= |z|^2|z'|^2$$

and the result follows upon taking square roots, since the numbers are non-negative. (19) follows from (18) as (17) did from (16).

Proof of (20) This is trivially true if $z + z' = 0$, so we need only consider the case when $z + z' \neq 0$. This means that $|z + z'|$ is also non-zero, so we have

$$\frac{|z| + |z'|}{|z + z'|} = \frac{|z|}{|z + z'|} + \frac{|z'|}{|z + z'|}$$

$$= \left|\frac{z}{z + z'}\right| + \left|\frac{z'}{z + z'}\right| \text{ by (19)}$$

$$\geq \text{Re}\left(\frac{z}{z + z'}\right) + \text{Re}\left(\frac{z'}{z + z'}\right) \text{ by (7)}$$

$$= \text{Re}\left(\frac{z + z'}{z + z'}\right) \text{ by (3)}$$

$$= 1$$

Multiplying both sides by the positive number $|z + z'|$ gives the required result.

The consequence of (13), (14), (16) and (17) is that we can compute the complex conjugate of any complex number, however complicated its expression, simply by replacing each occurrence of i by $-i$. For example, if

$$z = \left(i + \frac{(2 - 3i)(4 + 5i)^2}{2 + (i - (1 + i)^2)}\right)$$

then

$$\bar{z} = \left(-i + \frac{(2 + 3i)(4 - 5i)^2}{2 + (-i - (1 - i)^2)}\right)$$

EXERCISES 8.4.1

1 Find the value of

$$\left|\frac{(2 - 4i)(5 + i)}{(3 + i)(3 - 2i)}\right|.$$

2 Show that $|z + z'|^2 = |z|^2 + |z'|^2 + 2\,\text{Re}(z\bar{z}')$ and deduce that
$|z + z'|^2 + |z - z'|^2 = 2(|z|^2 + |z'|^2)$.

8.5 POLAR FORM OF COMPLEX NUMBERS

Let $z = x + iy$ be a non-zero complex number. The point P representing z in the Argand diagram has Cartesian coordinates (x, y); we now transform these to polar coordinates by writing $x = r \cos \theta$, $y = r \sin \theta$, where r is the distance of P from the origin, O, and θ is the angle, measured in an anti-clockwise direction, between the x axis and the line OP as shown in Figure 8.4.

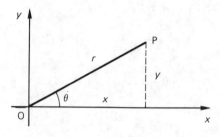

Fig. 8.4

In polar coordinates, z becomes

$$z = r(\cos\theta + i\sin\theta)$$

which is the *polar form* of z. Since $r^2 = x^2 + y^2 = |z|^2$, r is uniquely determined by z. The angle θ is called the *argument* of z and we write $\theta = \arg z$. Since we may add integer multiples of 2π to θ without changing the values of $\sin\theta$ and $\cos\theta$, θ is not uniquely determined by z. We therefore define a *principal value* of $\arg z$ by restricting θ to lie in the interval $(-\pi, \pi]$. It should be remembered that integer multiples of 2π can always be added to $\arg z$ if necessary.

Examples 8.5.1

z	r	$\arg z$ (principal value underlined)
1	1	$\ldots, -2\pi, \underline{0}, 2\pi, 4\pi, \ldots$
i	1	$\ldots, -3\pi/2, \underline{\dfrac{\pi}{2}}, \dfrac{5\pi}{2}, \dfrac{9\pi}{2}, \ldots$
$1+i$	$\sqrt{2}$	$\ldots, -\dfrac{7\pi}{4}, \underline{\dfrac{\pi}{4}}, \dfrac{9\pi}{4}, \dfrac{17\pi}{4}, \ldots$
-2	2	$\ldots, -\pi, \underline{\pi}, 3\pi, 5\pi, \ldots$
$-1+i$	$\sqrt{2}$	$\ldots, -\dfrac{5\pi}{4}, \underline{\dfrac{3\pi}{4}}, \dfrac{11\pi}{4}, \dfrac{19\pi}{4}, \ldots$

We use the last of these examples to illustrate a suitable method for finding the polar form of a complex number. First $r^2 = (-1)^2 + 1^2 = 2$, so $r = \sqrt{2}$, and we can write

$$-1 + i = \sqrt{2}\left(-\frac{1}{\sqrt{2}} + \frac{i}{\sqrt{2}}\right)$$

We therefore require the value of θ in $(-\pi, \pi)$ such that $\cos \theta = -\frac{1}{\sqrt{2}}$ and $\sin \theta = \frac{1}{\sqrt{2}}$. One way of evaluating θ is to find the angle whose tangent is -1. Both the angles $-\frac{\pi}{4}$ and $\frac{3\pi}{4}$ would satisfy this. However, we can discard the former, since this would give $\cos\left(-\frac{\pi}{4}\right) = \frac{1}{\sqrt{2}}$ and $\sin\left(-\frac{\pi}{4}\right) = -\frac{1}{\sqrt{2}}$, and hence $z = 1 - i$. The correct value of $-1 + i$ is obtained with $\theta = \frac{3\pi}{4}$.

The recommended procedure is therefore to use $\tan^{-1}\frac{y}{x}$ to find possible values for arg z, but to check which gives the correct sine and cosine values. It is usually a help to draw the given complex number in an Argand diagram, since this immediately gives an approximate answer, which can be used to check a more detailed calculation.

The rule for multiplying complex numbers becomes more natural when expressed in polar form. Let $z = r(\cos \theta + i \sin \theta)$ and $z' = r'(\cos \theta' + i \sin \theta')$. Then

$$zz' = rr'(\cos \theta + i \sin \theta)(\cos \theta' + i \sin \theta')$$
$$= rr'((\cos \theta \cos \theta' - \sin \theta \sin \theta')$$
$$+ i(\cos \theta \sin \theta' + \sin \theta \cos \theta'))$$
$$= rr'(\cos(\theta + \theta') + i \sin(\theta + \theta'))$$

where we have used standard trigonometric identities. Thus, to obtain the product of two complex numbers in polar form, we multiply the moduli and add the arguments.

Repeated applications of this rule allow us to extend it to the product of three or more numbers. For n complex numbers, we have, using an obvious notation,

$$z_1 z_2 \ldots z_n = (r_1 r_2 \ldots r_n)(\cos(\theta_1 + \theta_2 + \cdots + \theta_n)$$
$$+ i \sin(\theta_1 + \theta_2 + \cdots + \theta_n))$$

Letting $r_1 = r_2 = \cdots = r_n$ and $\theta_1 = \theta_2 = \cdots = \theta_n$ and writing the product of z

by itself n times as z^n, we obtain

$$z^n = r^n(\cos n\theta + i \sin n\theta).$$

This is *de Moivre's* Theorem for positive integer powers of a complex number. It provides a very easy way of evaluating powers. We now extend this to negative integer powers.

First, to find $1/z$, we proceed as follows:

$$\frac{1}{z} = \frac{1}{r(\cos\theta + i\sin\theta)} \cdot \frac{(\cos\theta - i\sin\theta)}{(\cos\theta - i\sin\theta)}$$

$$= \frac{1}{r}(\cos\theta - i\sin\theta)$$

$$= \frac{1}{r}(\cos(-\theta) + i\sin(-\theta))$$

We now define for non-zero z, $z^0 = 1$ and $z^{-n} = \dfrac{1}{z^n}$. The usual exponent rules $z^m z^n = z^{m+n}$ and $(z^m)^n = z^{mn}$ are then satisfied. We now easily combine the results for reciprocals and positive integer powers of a complex number to find for positive integers m

$$z^{-m} = \frac{1}{z^m}$$

$$= r^{-m}(\cos(-m\theta) + i\sin(-m\theta))$$

Finally, replacing $-m$ by n, we obtain the result for negative integer n

$$z^n = r^n (\cos n\theta + i \sin n\theta)$$

which extends de Moivre's Theorem to negative integer powers.

We can also use the reciprocal rule to obtain the quotient of $z' = r'(\cos\theta' + i\sin\theta')$ and $z = r(\cos\theta + i\sin\theta)$:

$$\frac{z'}{z} = z'\frac{1}{z}$$

$$= r'(\cos\theta' + i\sin\theta')\frac{1}{r}(\cos(-\theta) + i\sin(-\theta))$$

$$= \frac{r'}{r}(\cos(\theta' - \theta) + i\sin(\theta' - \theta))$$

which follows from the multiplication case above. The result shows that the quotient of two complex numbers in polar form is obtained from the quotient of the moduli and the difference of the arguments.

EXERCISES 8.5.1

1 Express the following complex numbers in polar form:

(i) $1 - i\sqrt{3}$; (ii) $-\sqrt{3} + i$; (iii) $-4i$; (iv) $\sin \phi + i \cos \phi$. ✗

2 Express $1 + i$ in polar form and hence evaluate $(1 + i)^4$ in the form $x + iy$.

8.6 EXPONENTIAL FORM OF COMPLEX NUMBERS

The complex number $\cos \theta + i \sin \theta$ is sometimes written as $e^{i\theta}$, where θ must be in radians. Some justification for this is provided by the polar form of product and de Moivre's Theorem, which in this notation become

$$e^{i(\theta + \theta')} = e^{i\theta}e^{i\theta'} \quad \text{and} \quad (e^{i\theta})^n = e^{in\theta}$$

These identities conform to the usual laws of exponents. Further justification comes from considering the infinite series (listed at the end of Chapter 2)

$$\sin x = x - \frac{x^3}{3!} + \frac{x^5}{5!} - \frac{x^7}{7!} + \cdots$$

$$\cos x = 1 - \frac{x^2}{2!} + \frac{x^4}{4!} - \frac{x^6}{6!} + \cdots$$

$$e^x = 1 + x + \frac{x^2}{2!} + \frac{x^3}{3!} + \frac{x^4}{4!} + \cdots$$

It can be shown (but not in this text) that these hold for all complex numbers x. The student should check, by putting $x = i\theta$, that $e^{i\theta} = \cos \theta + i \sin \theta$.

Using this notation, every non-zero complex number can be written in *exponential form* $re^{i\theta}$ with $r > 0$ unique and θ unique up to addition of integer multiples of 2π.

It is worth noting the remarkable formula obtained by taking $\theta = \pi$,

$$e^{i\pi} = -1$$

which relates the three important quantities e, i and π.

Finding nth Roots

Suppose we want to find the nth root of a non-zero complex number a, which is equivalent to solving $z^n = a$. Let $a = re^{i\theta}$ and $z = se^{i\beta}$. Then we have to solve $(se^{i\beta})^n = re^{i\theta}$, which becomes after expanding the left hand side $s^n e^{in\beta} = re^{i\theta}$.

Because the modulus and argument of a complex number are unique (apart from the addition of integer multiples of 2π in the case of the latter), we must have $s^n = r$ and $n\beta = \theta + 2k\pi$ for integer k. This gives

$$s = r^{1/n} \quad \text{and} \quad \beta = \frac{\theta + 2k\pi}{n}$$

and hence

$$z = r^{1/n} e^{i((\theta + 2k\pi)/n)}$$

The cases $k = 0, 1, \ldots, n-1$ give n distinct roots, since their arguments differ by at most $2(n-1)\dfrac{\pi}{n} < 2\pi$. Other values of k simply repeat these roots, since for any integer l

$$e^{i((\theta + 2(n+l)\pi)/n)} = e^{i((\theta + 2l\pi)/n + 2\pi)}$$
$$= e^{i((\theta + 2l\pi)/n)} e^{2i\pi}$$
$$= e^{i((\theta + 2l\pi)/n)}$$

since

$$e^{2\pi i} = \cos 2\pi + i \sin 2\pi = 1$$

Example 8.6.1

Find the cube roots of $8i$ and show them on an Argand diagram.

We first express $8i$ in polar form as $8e^{(i\pi/2)}$. The required roots are therefore

$$8^{1/3} e^{i(1/2 + 2k)(\pi/3)} \quad \text{with} \quad k = 0, 1, 2$$

that is,

$$2e^{(i\pi/6)}, \; 2e^{(5i\pi/6)} \quad \text{and} \quad 2e^{(3i\pi/2)}$$

The last of these has the value $-2i$ and the others can be evaluated using the equivalent polar form

$$2\left(\cos \frac{\pi}{6} + i \sin \frac{\pi}{6}\right) \quad \text{and} \quad 2\left(\cos \frac{5\pi}{6} + i \sin \frac{5\pi}{6}\right)$$

which yield the roots $\sqrt{3} + i$ and $-\sqrt{3} + i$. These are shown on an Argand diagram in Figure 8.5.

EXERCISES 8.6.1

1 Find the 4th roots of unity and indicate them on an Argand diagram.

2 Find the square roots of $\sqrt{3} + i$.

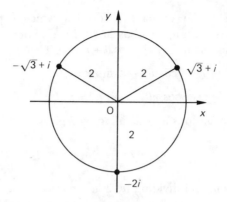

Fig. 8.5

8.7 TRIGONOMETRIC IDENTITIES

We can equate real and imaginary parts of complex identities (particularly de Moivre's Theorem) in polar form to obtain trigonometric identities.

Example 8.7.1

We expand the right-hand side of de Moivre's theorem with $n = 4$:

$$\cos 4\theta + i \sin 4\theta = (\cos \theta + i \sin \theta)^4$$

$$= \cos^4 \theta + 4 \cos^3 \theta (i \sin \theta) + 6 \cos^2 \theta (i \sin \theta)^2$$

$$+ 4 \cos \theta (i \sin \theta)^3 + (i \sin \theta)^4$$

$$= (\cos^4 \theta - 6 \cos^2 \theta \sin^2 \theta + \sin^4 \theta)$$

$$+ i(4 \cos^3 \theta \sin \theta - 4 \cos \theta \sin^3 \theta)$$

Equating real and imaginary parts, we obtain

$$\cos 4\theta = \cos^4 \theta - 6 \cos^2 \theta \sin^2 \theta + \sin^4 \theta$$

$$\sin 4\theta = 4 \cos^3 \theta \sin \theta - 4 \cos \theta \sin^3 \theta$$

We can also obtain an expression for $\tan 4\theta$ by dividing these, provided $\cos 4\theta \neq 0$:

$$\tan 4\theta = \frac{4 \cos^3 \theta \sin \theta - 4 \cos \theta \sin^3 \theta}{\cos^4 \theta - 6 \cos^2 \theta \sin^2 \theta + \sin^4 \theta}$$

$$= \frac{4 \tan \theta - 4 \tan^3 \theta}{1 - 6 \tan^2 \theta + \tan^4 \theta} \quad (\cos \theta \neq 0)$$

It is sometimes useful to have a formula which expresses products $\cos^m \theta \sin^n \theta$ in terms of $\cos \theta$, $\sin \theta$, $\cos 2\theta$, $\sin 2\theta$, ... For example, it provides a way of integrating them. Let $z = \cos \theta + i \sin \theta$. Then

$$\cos n\theta + i \sin n\theta = z^n$$

$$\cos n\theta - i \sin n\theta = z^{-n}$$

Adding these two equations and dividing by 2 gives

$$\cos n\theta = \frac{z^n + z^{-n}}{2} \tag{1}$$

and subtracting them and dividing by $2i$ gives

$$\sin n\theta = \frac{z^n - z^{-n}}{2i} \tag{2}$$

Example 8.7.2

Express $\cos^4 \theta \sin^3 \theta$ in terms of sines and cosines of multiples of θ.
  Using equations (1) and (2) with $n = 1$, we obtain

$$\cos^4 \theta \sin^3 \theta = (z + z^{-1})^4 (z - z^{-1})^3 / 24(2i)^3$$

$$= (z^2 - z^{-2})^3 (z + z^{-1})/24(2i)^3, \text{ since } (z + z^{-1})(z - z^{-1}) = z^2 - z^{-2},$$

$$= (z^6 - 3z^2 + 3z^{-2} - z^{-6})(z + z^{-1})/24(2i)^3$$

$$= (z^7 - z^{-7} - 3z^3 + 3z^{-3} + 3z^{-1} - 3z - z^{-5} + z^5)/24(2i)^3$$

$$= 2i(\sin 7\theta - 3 \sin 3\theta - 3 \sin \theta + \sin 5\theta)/24(2i)^3, \text{ using (1) and (2)},$$

$$= -(\sin 7\theta - 3 \sin 3\theta - 3 \sin \theta + \sin 5\theta)/64$$

EXERCISES 8.7.1

1 Find A, B, C and D such that

$$\cos^3 \theta \sin^5 \theta = (A \sin 8\theta + B \sin 6\theta + C \sin 4\theta + D \sin 2\theta)/128$$

2 Find a formula for $\tan 5\theta$ in terms of $\tan \theta$.

MISCELLANEOUS EXERCISES 8

1 A charged particle moving in a uniform magnetic field has equations of motion $\dfrac{d}{dt} v_x = \omega v_y$, $\dfrac{d}{dt} v_y = -\omega v_x$, where v_x, v_y are the x, y components of its velocity. Find a differential equation satisfied by $v_x + iv_y$ and show that it is satisfied by

$$v_x + iv_y = Ae^{-i\omega t}$$

where A is an arbitrary constant.

2 The solution of a certain differential equation is

$$x = Ae^{(\lambda + i\omega)t} + Be^{(\lambda - i\omega)t}$$

where A, B are arbitrary constants (which may be complex). Show that it may be written in the form

(a) $x = e^{\lambda t}(C \cos \omega t + D \sin \omega t)$, where $C = A + B$, $D = (A - B)i$
(b) $x = Ee^{\lambda t} \cos(\omega t + \phi)$, where $E = \sqrt{C^2 + D^2}$ and $\tan \phi = -(D/C)$

3 Show that $A \cos(\omega t + \phi)$ may be written as $\text{Re}(Ze^{i\omega t})$, where $Z = Ae^{i\phi}$ is the *phasor* of $A \cos(\omega t + \phi)$. Z contains information about the *amplitude* A and *phase* ϕ of the sinusoidal function.

4 Find the phasors of (i) $3 \cos\left(\omega t + \dfrac{\pi}{3}\right)$ and (ii) $-5 \cos\left(\omega t - \dfrac{\pi}{6}\right)$.

5 Let $Z_1 = A_1 e^{i\phi_1}$, $Z_2 = A_2 e^{i\phi_2}$. By considering $\text{Re}\{(Z_1 + Z_2)e^{i\omega t}\}$, show that the phasor of $A_1 \cos(\omega t + \phi_1) + A_2 \cos(\omega t + \phi_2)$ is $Z_1 + Z_2$.

6 Find the amplitude and phase of $3 \cos \omega t - 4 \sin \omega t$.

7 The voltage v and current c in a circuit of resistance R and inductance L are connected by the equation

$$v = Rc + L\frac{dc}{dt}$$

By writing $c = c_m e^{i\omega t}$, where c_m, ω are real constants, show that the voltage corresponding to a current $c_m \cos \omega t$ is $\text{Re}(Zc)$, where $Z = R + i\omega L$ is the *impedance* of the circuit.

8.8 ANSWERS TO EXERCISES

Exercises 8.2.1

1 $(2 - 3i) + (4 + 5i) = 6 + 2i, (2 - 3i)(4 + 5i) = 8 + 10i - 12i - 15i^2 = 23 - 2i$.

2 $2i - 9i^2 - (1 + 3i + i + 3i^2) - 5i = 11 - 7i$.

3 $i^{-1} = \dfrac{1}{i} \cdot \dfrac{-i}{-i} = -i$, $(1 + 4i)^{-1} = \dfrac{1}{1 + 4i} \cdot \dfrac{1 - 4i}{1 - 4i} = \dfrac{1 - 4i}{1^2 + 4^2} = \dfrac{1}{17} - \dfrac{4}{17}i$.

4 (i) $\dfrac{12 - 11i}{1 + 2i} \cdot \dfrac{1 - 2i}{1 - 2i} = \dfrac{-10 - 35i}{1^2 + 2^2} = -2 - 7i$,

(ii) $\dfrac{4+7i}{2+6i} \cdot \dfrac{2-6i}{2-6i} = \dfrac{50-10i}{2^2+6^2} = \dfrac{5}{4} - \dfrac{1}{4}i.$

Exercises 8.3.1

1   (i) Let $z = x + iy$, so $z^2 = x^2 - y^2 + 2xyi = -8 + 6i$, giving $x^2 - y^2 = -8$ (1)

and $2xy = 6$ (2). Solving (2) for y gives $y = \dfrac{3}{x}$, and substituting this

into (1) gives $x^2 - \dfrac{9}{x^2} + 8 = 0$, or $x^4 - 9x^2 + 8 = 0$, which is a quadratic

in x^2. Its roots are $x^2 = 1, -8$, but since x is real, we must have

$x = \pm 1$ and hence $y = \dfrac{3}{x} = \pm 3$, $z = \pm (1 + 3i)$. The answer should be

checked by squaring.

(ii) $z = \dfrac{i - (2 - i)}{1 + i} = \dfrac{-2 + 2i}{1 + i} \cdot \dfrac{1 - i}{1 - i} = 2i.$

(iii) With $z = x + iy$, the equation becomes $\sqrt{x^2 + y^2} - (x - iy) = 1 + 2i$,
giving $\sqrt{x^2 + y^2} - x = 1$, $y = 2$. Combining these and squaring, we
find $x^2 + 2^2 = (x + 1)^2 = x^2 + 2x + 1$, which gives $x = \frac{3}{2}$, $z = \frac{3}{2} + 2i$.
We *must* check this answer, since we squared both sides of an equation
to obtain it; we have $|\frac{3}{2} + 2i| - (\frac{3}{2} - 2i) = \sqrt{\frac{9}{4} + 4} - \frac{3}{2} + 2i = 1 + 2i$.

(iv) With $z = x + iy$, $x - iy = x^2 - y^2 + 2xyi$, so $x = x^2 - y^2$ (1) and
$y = -2xy$ (2). (2) gives either (a) $y = 0$, which yields from (1)
$x(1 - x) = 0$, hence $x = 0$ or $x = 1$, so $z = 0$ or 1, or (b) $x = -\frac{1}{2}$, which
yields from (1) $y^2 = \frac{3}{4}$, hence $z = -\frac{1}{2} \pm (\sqrt{3}/2)i$.

2   We simply verify the result:

$$z^2 = \dfrac{|a| + b}{2} - \dfrac{|a| - b}{2} - i\sqrt{|a|^2 - b^2} = b - i\sqrt{-c^2} = b - i(-c) = a,$$

since $\sqrt{-c^2} = -c$ when $c < 0$.

3   With $a = 1$, $b = -(1 + 4i)$, $c = -(3 - i)$, $b^2 - 4ac = -3 + 4i = (-1)(3 - 4i)$.
Thus $\sqrt{b^2 - 4ac} = \sqrt{-1}\sqrt{3 - 4i} = i\{\pm(2 - i)\} = \pm(1 + 2i)$ (from Example
8.3.1; alternatively the method used in that example can be used directly).
The solution of the quadratic is thus

$$z = \dfrac{(1 + 4i) \pm (1 + 2i)}{2} = 1 + 3i \text{ or } i.$$

4   Using the normal formula for z^2, we find

$$z^2 = \dfrac{2 \pm \sqrt{2^2 - 4 \times 4}}{2} = 1 \pm i\sqrt{3}.$$

The square root is found as in Example 8.3.1 as

$$\pm\left(\sqrt{\frac{3}{2}}\pm\frac{i}{\sqrt{2}}\right).$$

Exercises 8.4.1

1 The required value is

$$\sqrt{\frac{(2-4i)(2+4i)(5+i)(5-i)}{(3+i)(3-i)(3-2i)(3+2i)}}=\sqrt{\frac{20\times26}{10\times13}}=2.$$

2
$$|z+z'|^2 = (z+z')(\bar{z}+\bar{z}'), \text{ by (1) and (13)}$$
$$= z\bar{z} + z\bar{z}' + z'\bar{z} + z'\bar{z}'$$
$$= |z|^2 + |z'|^2 + z\bar{z}' + \overline{z\bar{z}}, \text{ by (1), (15) and (16)}$$
$$= |z|^2 + |z'|^2 + 2\operatorname{Re}(z\bar{z}') \text{ by (5)}.$$

Replacing z' by $-z'$ and using (12), we find $|z-z'|^2 = |z|^2 + |z'|^2 - 2\operatorname{Re}(z\bar{z}')$ and adding this to the first answer gives the second result.

Exercises 8.5.1

1 (i) $2\left(\cos\left(-\frac{\pi}{3}\right)+i\,\sin\left(-\frac{\pi}{3}\right)\right)$,

(ii) $2\left(\cos\left(\frac{5\pi}{6}\right)+i\,\sin\left(\frac{5\pi}{6}\right)\right)$,

(iii) $4\left(\cos\left(-\frac{\pi}{2}\right)+i\,\sin\left(-\frac{\pi}{2}\right)\right)$,

(iv) $\cos\left(\frac{\pi}{2}-\phi\right)+i\,\sin\left(\frac{\pi}{2}-\phi\right)$.

2
$$1+i=\sqrt{2}\left(\frac{1}{\sqrt{2}}+i\,\frac{1}{\sqrt{2}}\right)=\sqrt{2}\left(\cos\frac{\pi}{4}+i\,\sin\frac{\pi}{4}\right),$$

so

$$(1+i)^4=(\sqrt{2})^4(\cos\pi+i\,\sin\pi)=-4.$$

Exercises 8.6.1

1 $1 = e^{2k\pi i}$ with $k = 0, 1, 2, \ldots$. Then $1^{1/4} = e^{(k\pi i/2)} = \cos \dfrac{k\pi}{2} + i \sin \dfrac{k\pi}{2} = 1, i,$

$-1, -i$ for $k = 0, 1, 2, 3$.

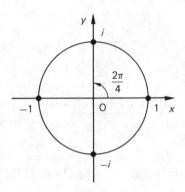

2 $\sqrt{3} + i = 2e^{(1/6 + 2k)\pi i}$ for $k = 0, 1, 2, \ldots$

$$\sqrt{\sqrt{3} + i} = \sqrt{2}e^{(1/12 + k)\pi i} = \pm\sqrt{2}\left(\cos \frac{\pi}{12} + i \sin \frac{\pi}{12}\right)$$

Exercises 8.7.1

1

$$\cos^3 \theta \sin^5 \theta = (z + z^{-1})^3 2^{-3}(z - z^{-1})^5 (2i)^{-5}$$

$$= (z^2 - z^{-2})^3 (z - z^{-1})^2 2^{-8} \frac{1}{i}$$

$$= (z^6 - 3z^2 + 3z^{-2} - z^{-6})(z^2 - 2 + z^{-2})2^{-8} \frac{1}{i}$$

$$= \{(z^8 - z^{-8}) - 2(z^6 - z^{-6}) - 2(z^4 - z^{-4})$$

$$+ 6(z^2 - z^{-2})\}2^{-8} \frac{1}{i}$$

$$= \frac{1}{128} (\sin 8\theta - 2 \sin 6\theta - 2 \sin 4\theta + 6 \sin 2\theta).$$

2 $\cos 5\theta + i \sin 5\theta = (\cos \theta + i \sin \theta)^5$

$$= \cos^5 \theta + 5i \cos^4 \theta \sin \theta - 10 \cos^3 \theta \sin^2 \theta$$

$$- 10i \cos^2 \theta \sin^3 \theta + 5 \cos \theta \sin^4 \theta + i \sin^5 \theta.$$

Equating real and imaginary parts we find

$$\cos 5\theta = \cos^5 \theta - 10 \cos^3 \theta \sin^2 \theta + 5 \cos \theta \sin^4 \theta,$$

$$\sin 5\theta = 5 \cos^4 \theta \sin \theta - 10 \cos^2 \theta \sin^3 \theta + \sin^5 \theta,$$

so

$$\tan 5\theta = \frac{5 \cos^4 \theta \sin \theta - 10 \cos^2 \theta \sin^3 \theta + \sin^5 \theta}{\cos^5 \theta - 10 \cos^3 \theta \sin^2 \theta + 5 \cos \theta \sin^4 \theta}$$

$$= \frac{5 \tan \theta - 10 \tan^3 \theta + \tan^5 \theta}{1 - 10 \tan^2 \theta + 5 \tan^4 \theta}$$

provided $\cos \theta \neq 0$ and $\cos 5\theta \neq 0$.

Miscellaneous Exercises 8

1
$$\frac{d}{dt}(v_x + iv_y) = \omega(v_y - iv_x) = -i\omega(v_x + iv_y).$$

The general solution of this differential equation is $v_x + iv_y = Ae^{-i\omega t}$.

2 (a) $x = e^{\lambda t}(Ae^{i\omega t} + Be^{-i\omega t})$
$$= e^{\lambda t}\{A(\cos \omega t + i \sin \omega t) + B(\cos \omega t - i \sin \omega t)\}$$
$$= e^{\lambda t}(C \cos \omega t + D \sin \omega t)$$

with C, D as given.

(b) $Ee^{\lambda t} \cos(\omega t + \phi) = e^{\lambda t}(E \cos \phi \cos \omega t - E \sin \phi \sin \omega t) = x$,

if $E \cos \phi = C$, $E \sin \phi = -D$, i.e. if $E = \sqrt{C^2 + D^2}$, $\tan \phi = -\dfrac{D}{C}$.

3 $\operatorname{Re}(Ze^{i\omega t}) = \operatorname{Re}(Ae^{i(\omega t + \phi)}) = A \cos(\omega t + \phi)$.

4 (i) $3e^{(i\pi/3)}$,

(ii) since $-5 \cos\left(\omega t - \dfrac{\pi}{6}\right) = 5 \cos\left(\pi - \omega t + \dfrac{\pi}{6}\right) = 5 \cos\left(\omega t - \dfrac{7\pi}{6}\right)$,

the phasor is $5e^{-(7\pi/6)}$.

5 $\operatorname{Re}\{(Z_1 + Z_2)e^{i\omega t}\} = \operatorname{Re}\{A_1 e^{i(\omega t + \phi_1)} + A_2 e^{i(\omega t + \phi_2)}\}$

$$= A_1 \cos(\omega t + \phi_1) + A_2 \cos(\omega t + \phi_2), \text{ so } Z_1 + Z_2$$

is the phasor of the last expression.

6 $3 \cos \omega t - 4 \sin \omega t = 5 \cos(\omega t + \phi)$. This has amplitude 5 and phase $\phi = \tan^{-1} \frac{4}{3}$.

7 With $c = c_m e^{i\omega t}$, $v = Rc + L\dfrac{dc}{dt} = (R + i\omega L)c_m e^{i\omega t} = Zc$. The voltage corresponding to $c_m \cos \omega t = \operatorname{Re}(c_m e^{i\omega t})$ is then $\operatorname{Re}(Zc)$.

9 DIFFERENTIAL EQUATIONS

9.1 INTRODUCTION

Students coming across the topic for the first time often have difficulty in appreciating just what a differential equation is. We shall postpone discussion of this problem until we have looked at some simple examples, which also serve to give motivation for the solution of differential equations.

Examples 9.1.1

1. Consider a car under test, whose acceleration is a measured function of time. Let us suppose that the acceleration, a, is well-approximated by the formula

$$a = 1 - 2t$$

where t is the time. Letting v be the speed of the car, we may replace a by $\mathrm{d}v/\mathrm{d}t$ to obtain the differential equation

$$\frac{\mathrm{d}v}{\mathrm{d}t} = 1 - 2t$$

We should like to obtain a 'solution' of this differential equation: that is, we should like to find a function v of time, which, when substituted into the differential equation, satisfies it. We can solve this simple example directly by integrating with respect to t to obtain

$$v = \int (1 - 2t)\,\mathrm{d}t$$

$$= t - t^2 + c \tag{1}$$

This does not give a specific solution, because the constant of integration is quite arbitrary. Whatever value we take for it, the differential equation

270

is satisfied, since it disappears as soon as we differentiate v. Suppose, however, we have the additional information that the car started from rest at time zero. Putting $v = 0$ and $t = 0$ into Equation (1), we find $c = 0$, so the required solution is

$$v = t - t^2$$

2. We now take an example from economics. Financial investments normally acquire interest in finite increments (for example, once a month). We shall model this situation by assuming that the interest accrues continuously according to the rule

$$\begin{array}{ccc} \text{rate of increase} & = & \text{present value} \\ \text{of investment} & & \text{of investment} \end{array} \times \begin{array}{c} \text{rate of interest} \\ \text{per unit investment} \end{array}$$

or

$$\frac{dP}{dt} \quad = \quad P \quad \times \quad r \quad \quad (1)$$

Here P, the current value of the investment, varies with time, t, and r is a constant. For example, if the interest rate is 10% per annum and t is in years then $r = 0.1$.

We cannot proceed as before, since we should arrive at

$$P = \int Pr \, dt$$

and the integral cannot be evaluated, since we do not know what function of t P is until we have solved the problem. For the moment, we try an *ad hoc* approach. To simplify the problem, we set $r = 1$ to obtain

$$\frac{dP}{dt} = P \quad \quad (2)$$

We now ask what function, when differentiated, gives the same function. The answer of course is the exponential function, which in these variables means that $P = e^t$. We can see that this does indeed satisfy Equation (2) by differentiating $P = e^t$ to obtain

$$\frac{dP}{dt} = e^t = P$$

If we restore r to the differential equation, we soon see that this changes the solution to e^{rt}; an arbitrary constant k can also be inserted, so that finally we obtain the solution of Equation (1) as

$$P = ke^{rt} \quad \quad (3)$$

The student should check that this does satisfy Equation (1) by differentiation.

If we are given that $P = P_0$ when $t = 0$, we substitute these values into Equation (3) to find $P_0 = ke^0$, so that $k = P_0$ and the required solution is

$$P = P_0 e^{rt}$$

This example is important because it models many real-life situations; it represents exponential growth if $r > 0$ (for example, population growth), or exponential decay if $r < 0$ (for example, radioactive decay).

Although we have managed to find solutions for these two examples, the methods used, especially for the second, have been rather arbitrary. We shall shortly be looking at systematic methods of solution, but for the moment we look at two further examples to show the relation between differential equations and their solutions, as a preliminary to defining these terms.

Examples 9.1.2

1. Consider the line of slope 2 through the origin, whose equation is $y = 2x$. Differentiation gives the differential equation

$$\frac{dy}{dx} = 2$$

Integrating in order to obtain the solution, we obtain

$$y = 2x + c$$

We now regard the arbitrary constant of integration, c, as a *parameter*. When $c = 0$, we obtain the original line; $c = 1$ gives a line, also of slope 2, but through the point $(0, 1)$. Each value of c gives a different line; we call the set of these for all values of c the *family of solution curves* of the differential equation. Some members are shown in Figure 9.1. All have

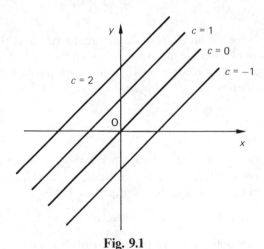

Fig. 9.1

the same slope of 2, and this is precisely the information contained in the differential equation. Each member solves the differential equation, that is, satisfies it; for differentiating $y = 2x + c$ gives $dy/dx = 2$ regardless of the value of c.

2. Consider the family of curves given by

$$y = ce^{2x} \tag{1}$$

where c is a parameter. Some members are shown in Figure 9.2. Differentiation gives

$$\frac{dy}{dx} = 2ce^{2x} \tag{2}$$

This looks like our previous examples of differential equations, except that it contains the parameter c. We now require that the same differential equation be satisfied on *every* solution curve; we achieve this by eliminating c from Equations (1) and (2) to obtain

$$\frac{dy}{dx} = 2(ce^{2x}) = 2y$$

The differential equation, whose solution curves are given by Equation (1), is therefore

$$\frac{dy}{dx} = 2y \tag{3}$$

This again gives the slope of any solution curve at any point on it, provided we put in the value of y for the point.

We can pick out a particular member of Equation (1) by requiring it to pass through a given point. The specification of such a point is called

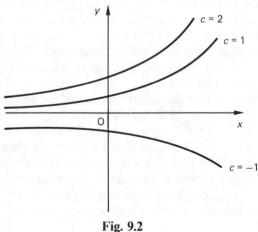

Fig. 9.2

273

an *initial condition*. In this example, the initial condition $y = -2$ when $x = 0$ substituted into Equation (1) gives $-2 = ce^0$ and hence $c = -2$. The required solution is then

$$y = -2e^{2x}$$

We now consider whether the family (1) contains all possible solutions of the differential equation (3). Writing this in the form

$$y' - 2y = 0$$

and multiplying through by e^{-2x}, we obtain

$$y'e^{-2x} - 2ye^{-2x} = 0$$

This can be written (using implicit differentiation) as

$$\frac{d}{dx}(ye^{-2x}) = 0$$

which integrates to

$$ye^{-2x} = c$$

from which we recover Equation (1) by multiplying through by e^{2x}. Thus, Equation (1) includes all possible solutions of the differential Equation (3). We also deduce that the full solution must contain an arbitrary constant, since an integration is required to obtain the solution. Such a solution, which contains an arbitrary constant, is called the *general solution* of the differential equation.

The examples we have so far looked at have only contained first order derivatives, but higher order derivatives may occur, as we shall see later.

We now define what we mean by a differential equation and its solution. For ease of description, we shall use the variables x and y, but any other two variables would do just as well.

A *differential equation* is an equation connecting, x, y and derivatives of y with respect to x. The *order* of a differential equation is the order of the highest derivative which occurs.

A *solution* of a differential equation is any function of x which, when substituted for y, satisfies the differential equation.

The existence of solutions for reasonably well-behaved differential equations has been proved. We shall assume that a solution always exists, so that we may regard y as a function of x whenever we have a differential equation in the variables x and y. This enables us to abbreviate the initial condition $y = y_0$ when $x = x_0$ as $y(x_0) = y_0$. The next four sections provide a systematic approach to the solution of some of the simpler types of first-order differential equation.

EXERCISE 9.1.1

Sketch the family of curves $y = ce^{-2x}$ for different values of c and find the differential equation satisfied along each member of the family.

9.2 DIFFERENTIAL EQUATIONS OF TYPE 1 (SEPARABLE)

We split Type 1 into two subtypes, 1a and 1b.

Type 1a

This has the general form

$$\frac{dy}{dx} = f(x)$$

where f is a function of x alone. Since $f(x)$ is the derivative of y, y is obtained by integrating $f(x)$ with respect to x, giving as the required solution

$$y = \int f(x)\,dx$$

Examples 9.2.1

1.
$$\frac{dy}{dx} = \sin x$$

This is clearly an example of Type 1a. Thus,

$$y = \int \sin x\,dx$$

$$= -\cos x + c$$

is the general solution.

Suppose that we are given the initial condition

$$y\left(\frac{\pi}{2}\right) = 2$$

Substituting this into the general solution, we obtain

$$2 = -\cos\frac{\pi}{2} + c$$

giving $c = 2$ and the required solution as

$$y = 2 - \cos x$$

2. $$(1+x)\frac{dy}{dx} = 1$$

This is not in the standard form of Type 1a; however, dividing through by $(1+x)$ gives

$$\frac{dy}{dx} = 1/(1+x)$$

which is of Type 1a. Thus

$$y = \int (1+x)^{-1}\, dx = \ln(1+x) + \ln k$$

$$= \ln k(1+x)$$

Note that we took the arbitrary constant of integration in the form $\ln k$ so that it combined neatly with the other log term. This becomes useful in later problems because it allows easier simplification.

Before proceeding to the next type, it is useful to arrange the solution of $dy/dx = f(x)$ slightly differently. Although we know that dy/dx is not a fraction, in which dy and dx are separate numbers, we nevertheless manipulate it as a fraction for the present purpose. The justification is that we obtain the same expression for the answer as in the previous method. Thus, we multiply both sides of $dy/dx = f(x)$ by dx to obtain

$$dy = f(x)\, dx \qquad (1)$$

and then insert integral signs to obtain

$$\int dy = \int f(x)\, dx \qquad (2)$$

and since $\int dy = y$, we rederive

$$y = \int f(x)\, dx$$

In future we shall omit line (1) by inserting the integral signs immediately to obtain line (2).

Type 1b

This has the general form

$$\frac{dy}{dx} = g(y)$$

Here the right-hand side is a function of y alone and we cannot integrate it with respect to x. However, since $dy/dx = 1/(dx/dy)$, we find

$$\frac{1}{g(y)} = \frac{1}{\dfrac{dy}{dx}} = \frac{dx}{dy}$$

This now has the form of Type 1a with the roles of x and y interchanged. Multiplying both sides by dy and inserting integral signs, we obtain

$$\int \frac{dy}{g(y)} = \int dx = x + c \tag{1}$$

We can again obtain the solution more directly by regarding dy/dx as a fraction. Thus, multiplying $dy/dx = g(y)$ through by $dx/g(y)$ and inserting integral signs gives Equation (1) again.

Examples 9.2.2

1.
$$\frac{dy}{dx} = -y^2$$

This is clearly of Type 1b, so if $y \neq 0$ we can multiply through by $-dx/y^2$; inserting integral signs, we obtain

$$-\int y^{-2}\,dy = \int dx$$

this integrates to

$$\frac{1}{y} = x + c$$

which upon rearranging becomes

$$y = \frac{1}{x + c}$$

Note that we were only able to carry through the solution by assuming that $y \neq 0$. However, if $y = 0$ then $dy/dx = 0$ also, and so $y = 0$ is a solution of the differential equation. This is obtained from the general solution by letting c go to infinity.

2.
$$\frac{dP}{dt} = rP$$

where r is a constant. This is the example from Economics which we solved earlier in a rather *ad hoc* manner. Now we recognise it as being of Type 1b, so we multiply through by dt/P and insert integral signs to obtain

$$\int \frac{\mathrm{d}P}{P} = \int r \, \mathrm{d}t$$

which integrates to

$$\ln P = rt + c$$

Exponentiating, we obtain

$$P = e^{rt+c}$$

$$= e^{rt} \cdot e^{c}$$

$$= k e^{rt}$$

where $k = e^{c}$ is an alternative form of arbitrary constant.

Type 1 (Separable Variables)

This type is a combination of Types 1a and 1b and takes the general form

$$\frac{\mathrm{d}y}{\mathrm{d}x} = \frac{f(x)}{g(y)}$$

Although the right-hand side depends on both x and y, it is of such a form that the parts dependent on x can be separated from the parts that depend on y. Multiplying through by $g(y)\,\mathrm{d}x$ and inserting the integral signs gives

$$\int g(y)\,\mathrm{d}y = \int f(x)\,\mathrm{d}x$$

and we have reduced the problem to that of two integrations.

Examples 9.2.3

1. $$\frac{\mathrm{d}y}{\mathrm{d}x} = -2xy$$

The right-hand side is clearly separable, so multiplying through by $\mathrm{d}x/y$ and inserting integral signs, we obtain

$$\int y^{-1}\,\mathrm{d}y = \int (-2x)\,\mathrm{d}x$$

So

$$\ln y = -x^2 + c$$

giving

$$y = e^{-x^2 + c}$$

$$= ke^{-x^2}$$

where k replaces e^c.

2.
$$\frac{dy}{dx} = \frac{x}{y}$$

Multiplying through by $y\,dx$ and inserting integral signs gives

$$\int y\,dy = \int x\,dx$$

Then

$$\tfrac{1}{2}y^2 = \tfrac{1}{2}x^2 + \tfrac{1}{2}c$$

where the arbitrary constant has been written as $\tfrac{1}{2}c$ so that we may conveniently multiply through by 2 to obtain

$$y^2 = x^2 + c$$

Taking square roots, we obtain the general solution

$$y = \pm\sqrt{x^2 + c}$$

The alternative signs look a little confusing, but remember that the general solution contains all solutions. Some of these are shown in Figure 9.3.

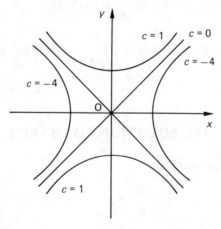

Fig. 9.3

A given initial condition will pick out just one curve. For example, $y(0) = 1$ substituted into the general solution gives us

$$1 = \pm\sqrt{0 + c}$$

so that c takes the value 1. This appears to give the alternative solutions

$$y = + \sqrt{x^2 + 1} \quad \text{or} \quad y = -\sqrt{x^2 + 1}$$

We pick the first of these solutions to avoid a discontinuous jump from the point $(0, 1)$, corresponding to the initial condition, onto the second solution, which starts at $(0, -1)$.

EXERCISES 9.2.1

1 Solve the following Type 1a differential equations:

(i) $\dfrac{dy}{dx} = \cos x;$ (ii) $x\dfrac{dy}{dx} = 2;$ (iii) $x\dfrac{dy}{dx} + x + 1 = 0.$

2 Find the general solutions of the following Type 1b differential equations:

(i) $\dfrac{dy}{dx} = e^{-y};$ (ii) $\dfrac{dy}{dx} = 1/2y;$

(iii) $(1 + y)\dfrac{dy}{dx} = y;$ (iv) $\dfrac{dx}{dt} = 1/2x.$

3 Solve the following Type 1 differential equations:

(i) $\dfrac{dy}{dx} = xy \quad (= x/y^{-1});$ (ii) $\dfrac{dy}{dx} = y^2 \sin x;$

(iii) $(x + 1)\dfrac{dy}{dx} = y;$ (iv) $(1/y)\dfrac{dy}{dx} = (x + 2)/(x^2 + x).$

9.3 DIFFERENTIAL EQUATIONS OF TYPE 2

This type has the general form

$$\frac{dy}{dx} = f(y/x) \tag{1}$$

in which x and y only occur in the combination y/x. We convert this to separable form by making the substitution $y = vx$. We find dy/dx by differentiating $y = vx$ to obtain

$$\frac{dy}{dx} = v + x\frac{dv}{dx}$$

and substituting into Equation (1), we obtain

$$v + x\frac{dv}{dx} = f(v)$$

which is a Type 1 equation in the new variable, v, and x.

Example 9.3.1

$$\frac{dy}{dx} = \frac{x+y}{x} \tag{1}$$

The right-hand side may be written $1 + y/x$ and is hence of the form $f(y/x)$. Putting $y = vx$, Equation (1) becomes

$$v + x\frac{dv}{dx} = 1 + v$$

so

$$x\frac{dv}{dx} = 1$$

and multiplying through by dx/x, we obtain

$$\int dv = \int x^{-1}\, dx$$

which integrates to

$$v = \ln x + \ln k$$
$$= \ln kx$$

Finally, putting $v = y/x$, we obtain

$$y = x \ln kx$$

EXERCISE 9.3.1

Find the general solution of the differential equation

$$y' = \frac{x^2 + xy + y^2}{x^2}, \quad x > 0$$

9.4 DIFFERENTIAL EQUATIONS OF TYPE 3 (EXACT)

In order to introduce this type, we differentiate implicitly with respect to x the equation

$$x^2 + y^2 = 0$$

to obtain the differential equation

$$2x + 2yy' = 0$$

To solve this, we just reverse the procedure; first we write it as

$$\frac{d}{dx}(x^2 + y^2) = 0$$

when it is clear that it can be integrated to

$$x^2 + y^2 = c$$

A differential equation which can be obtained by implicit differentiation of a function of x and y is called *exact*. The general form is

$$\frac{d}{dx}f(x, y) = 0 \qquad (1)$$

and its solution is

$$f(x, y) = c$$

The problem is to recognise when we have the form (1). We start by carrying out the implicit differentiation in Equation (1), which gives

$$f_x(x, y) + f_y(x, y)y' = 0 \qquad (2)$$

Now suppose that the equation we wish to solve has the form

$$p(x, y) + q(x, y)y' = 0 \qquad (3)$$

Equations (2) and (3) are the same if

$$p(x, y) = f_x(x, y) \quad \text{and} \quad q(x, y) = f_y(x, y)$$

If we differentiate the first of these partially with respect to y and the second with respect to x, we obtain

$$p_y(x, y) = f_{xy}(x, y) \quad \text{and} \quad q_x(x, y) = f_{yx}(x, y).$$

But for a well-behaved function, we know that $f_{xy} = f_{yx}$, so that we must have $p_y = q_x$. The converse of this, that if $p_y = q_x$ then the differential equation (3) is exact, can also be proved. We can thus check whether a given differential equation is exact.

The reader may have noticed the similarity of this development to that for the line integral of a conservative vector field in Section 7.4. f can be regarded as a potential, and can be found in a similar way (see alternative solution of Example 7.4.2.2). However, it is usually easy to spot the solution directly, as in the following example.

Example 9.4.1

$$xy' + y - 2x = 0$$

Comparing this with Equation (3), we have $p(x, y) = y - 2x$ and hence $p_y(x, y) = 1$ and $q(x, y) = x$ hence, $q_x(x, y) = 1$. The equation is thus exact and can be written as

$$\frac{d}{dx}(xy - x^2) = 0$$

which has the solution

$$xy - x^2 = c$$

Exact equations do not commonly occur in practice, but they provide a useful tool for deriving a method of solution of Type 4 equations.

EXERCISE 9.4.1

Check that the following differential equations are exact and integrate them:

(i) $3x^2 - (xy' + y) + 2yy' = 0$;
(ii) $y' \cos x - y \sin x = 0$.

9.5 DIFFERENTIAL EQUATIONS OF TYPE 4 (LINEAR)

A differential equation is *linear* if it is linear in y and y', that is if each term is of degree 1 in y or y' (but *not* both) or contains neither. For example

$$y' + y = x^2 \quad \text{and} \quad y' + \frac{y}{\sin x} = \cos x$$

are both linear, while

$$y' + \sin y = x, \; y' + y^2 = 0 \text{ and } (y')^2 + y = x$$

are not linear. Note that the way that x appears does not affect the linearity of the equation.

Before deriving a general method of solving linear equations, we look at an example which shows the idea. The differential equation

$$y' + \frac{y}{x} = 2$$

is clearly linear, but not exact. Multiplying through by x, however, gives

$$xy' + y = 2x$$

which is the example we gave earlier of an exact equation.

In this example we made the differential equation exact by multiplying through by x. We call x an *integrating factor*, and we now derive a method of finding an integrating factor for the general form of linear differential equation, which we shall take as

$$y' + y \cdot p(x) = q(x) \tag{1}$$

where p and q are functions of x only. Suppose that $f(x)$ is an integrating factor; multiplying Equation (1) through by this gives

$$y' \cdot f(x) + y \cdot p(x) \cdot f(x) = q(x) \cdot f(x) \tag{2}$$

Now compare this with

$$y' \cdot f(x) + y \cdot f'(x) = \frac{d}{dx}(y \cdot f(x)) \tag{3}$$

If we can choose the function f in such a way that

$$f'(x) = p(x) \cdot f(x) \tag{4}$$

then the left-hand side of Equations (2) and (3) will be the same and hence the right-hand sides also, so that

$$\frac{d}{dx}(y \cdot f(x)) = q(x) \cdot f(x) \tag{5}$$

Equation (5) is in a form which we can integrate directly. We must now find the function f from Equation (4), which we first rearrange and insert integral signs to obtain

$$\int (f'(x)/f(x)) \, dx = \int p(x) \, dx$$

Making the substitution $t = f(x)$, $dt = f'(x) \, dx$ in the left-hand side, we obtain

$$\int \frac{dt}{t} = \int p(x) \, dx$$

which gives

$$\ln t = \int p(x) \, dx$$

so that

$$t = f(x) = e^{\int p(x)\,dx}$$

and this is our desired integrating factor.

Note that, when evaluating the integral for f, we do not need to add an arbitrary constant. For suppose we added k; this would multiply the integrating factor by e^k, a constant, and all this would do is multiply the whole differential equation by this constant, with no effect on the solution.

The following steps are recommended for solving examples of linear differential equations:

(1) Rearrange if necessary and compare with Equation (1).
(2) Evaluate the integrating factor, $e^{\int p(x)\,dx}$.
(3) Multiply through by the integrating factor.
(4) Write the left-hand side as $d/dx\,(\dots)$.
(5) Integrate.

Examples 9.5.1

$$y' = x + \frac{2y}{x}$$

1. This example is clearly linear, so following the recommended steps, we obtain successively

 1. $y' + y \cdot (-2/x) = x$ (by rearranging)
    $y' + y \cdot p(x) = q(x)$ (for comparison)
    so $p(x) = -2/x$ (note the minus sign; it is essential)
    and $q(x) = x$
 2. $e^{\int p(x)\,dx} = e^{-\int 2\,dx/x} = e^{-2\ln|x|} = e^{\ln x^{-2}} = x^{-2}$, where we have omitted the modulus signs, since x occurs as a square.
 3. $y'x^{-2} - 2yx^{-3} = x^{-1}$ (multiplying by the integrating factor)
 4. $d/dx(yx^{-2}) = x^{-1}$
 5. $yx^{-2} = \int x^{-1}\,dx = \ln|x| + \ln|k| = \ln|kx|$
    giving

$$y = x^2 \ln|kx|$$

Since we have performed a number of steps to solve this problem, it is worth checking our solution. To do this, differentiate it to obtain

$$y' = 2x \ln|kx| + x^2 \frac{1}{x}$$

285

Eliminating k between this and the solution, we obtain

$$y' = \frac{2y}{x} + x$$

2. Consider a descending parachutist. It is found by observation that the resisting force is roughly proportional to the speed. Suppose that the speed downwards is v at time t. Then the equation of motion will be

$$mv' = mg - kv \quad\quad\quad (1)$$

where v' here means the derivative of v with respect to t, g is the acceleration due to gravity, m is the mass of the parachutist and k is a positive constant. This differential equation is linear and we solve it using the recommended steps.

(1) $v' + \dfrac{k}{m}v = g$

(2) $e^{\int p(t)\,dt} = e^{\int k/m\,dt} = e^{kt/m}$ (note use of the variable t in place of x)

(3) $v'e^{kt/m} + (k/m)ve^{kt/m} = ge^{kt/m}$

(4) $\dfrac{d}{dt}(ve^{kt/m}) = ge^{kt/m}$

(5) $ve^{kt/m} = \dfrac{mg}{k}e^{kt/m} + c$

Multiplying through by $e^{-kt/m}$ (taking care to remember that c has to be multiplied by this factor!) we obtain the solution as

$$v = mg/k + ce^{-kt/m}$$

We note that, if we let $t \to \infty$, the second term goes to zero, and we obtain the 'terminal velocity' $v = mg/k$; thus, eventually the speed of fall becomes essentially constant.

Suggested Procedure for Examples of Unspecified Type

The problem is to identify which type the example belongs to. This entails comparison with each of the standard types in turn. It is best to start with the most easily recognisable, then the next easiest, and so on. This suggests the order

First: Type 4 (linear, $y' + yp(x) = q(x)$)
Second: Type 2 ($y' = f(y/x)$)
Third: Type 1 (separable, $y' = f(x)/g(y)$)
Last: Type 3 (exact, $(d/dx)f(x, y) = 0$)

Sometimes, the given differential equation has to be rearranged, or perhaps its variables changed, in order to be recognised as one of these types. Once recognised, the standard method of solution for the appropriate type should be used.

Some differential equations belong to more than one type; this does not matter, since any one of the methods will give the solution.

EXERCISES 9.5.1

1 Solve the following linear differential equations:

(i) $y' + 2y = 1$;   (ii) $y' - \dfrac{2y}{x} = x^2$;   (iii) $y' + \dfrac{y}{x} = \dfrac{1}{x}$;

(iv) $y' - 2xy = e^{x^2}$;   (v) $y' = y + 1$.

2 Specify which type the following differential equations are:

(i) $y' + 6xy = x^2$;   (ii) $(\cos x)(\sin y) + (\sin x)(\cos y)y' = 1$;

(iii) $\dfrac{1}{y} - \dfrac{xy'}{y^2} = 2$;   (iv) $y' = 4x^2y^2$;

(v) $(\cos x)y' = xy^2$.

9.6 AN ALTERNATIVE WAY OF SOLVING FIRST-ORDER LINEAR DIFFERENTIAL EQUATIONS

Why should we need another method, since we already have a perfectly good one? The reason is that the method we are going to describe will work equally well for higher-order equations, but it is easier to introduce for first-order equations.

We split the solution of

$$y' + yp(x) = q(x) \tag{1}$$

into two parts.

(a) Solve the equation obtained from Equation (1) by making the right-hand side zero,

$$y' + yp(x) = 0 \tag{2}$$

This is a separable equation; suppose that $f(x)$ is a solution, so that, substituting into Equation (2), we must have

$$f'(x) + f(x)p(x) = 0 \tag{3}$$

(b) Now suppose that we know *any* solution of Equation (1), $y = g(x)$ say, so that we must have

$$g'(x) + g(x)p(x) = q(x) \qquad (4)$$

Then we assert that the general solution of Equation (1) is

$$y = Af(x) + g(x) \qquad (5)$$

where A is an arbitrary constant. We call $Af(x)$ the *complementary function* and $g(x)$ a *particular integral* of the differential equation (1). The general solution is thus the sum of the complementary function and a particular integral.

To show this, differentiate Equation (5), to find y', and substitute this and y from Equation (5) into the left-hand side of Equation (1), to obtain

$$y' + yp(x) = Af'(x) + g'(x) + (Af(x) + g(x))p(x)$$
$$= A(f'(x) + f(x)p(x)) + (g'(x) + g(x)p(x))$$
$$= q(x)$$

where we have used Equations (3) and (4) to obtain the last line. This shows that Equation (5) gives a solution of Equation (1); we must now show that it contains all solutions.

Suppose we are given the initial condition $y(a) = y_a$; substituting this into Equation (5) gives

$$y_a = Af(a) + g(a)$$

and so

$$A = \frac{y_a - g(a)}{f(a)} \quad \text{if} \quad f(a) \neq 0$$

Thus, provided $f(a) \neq 0$, we can find a value of A for any initial condition, so that all solutions are contained in Equation (5).

The remaining problem is to find a particular integral; we give a few examples before suggesting a standard procedure.

Examples 9.6.1

1.
$$y' - 2y = x \qquad (1)$$

(a) For the complementary function, we must solve

$$y' - 2y = 0$$

This is linear (or separable), with integrating factor e^{-2x}. After multiplying through by this, we can write the differential equation as $d/dx(ye^{-2x}) = 0$, which gives after integration, $ye^{-2x} = A$, and so the solution is

$$y = Ae^{2x}$$

(b) How do we find a particular integral? Without foreknowledge, it seems as though we must guess a suitable form of solution. After a few examples however, it will become clear what is required.

In this example, the right-hand side of Equation (1) is just x, so we look for a form for y such that the left hand side, $y' - 2y$, equals x. We try $y = ax + b$, where a and b are constants. Then $y' = a$ and, substituting into Equation (1), we obtain

$$y' - 2y = a - 2(ax + b) = x$$

or

$$(a - 2b) - 2ax = x$$

We want to find values of a and b so that this equation is satisfied for all values of x. The coefficient of x and the constant term must therefore be the same on both sides of the equation, so that $-2a = 1$ and $a - 2b = 0$, giving $a = -\frac{1}{2}$ and $b = -\frac{1}{4}$. Thus, a particular integral is

$$y = -\frac{x}{2} - \frac{1}{4}$$

and the general solution is

$$y = Ae^{2x} - \frac{x}{2} - \frac{1}{4}$$

2. $$y' - 2y = e^{3x}$$

(a) The left-hand side is the same as that in the previous example, so the complementary function is the same.

(b) For a particular integral, we need y to be such that $y' - 2y$ equals e^{3x}. This suggests that we try $y = ae^{3x}$, where a is a constant. Putting this into the differential equation gives

$$y' - 2y = 3ae^{3x} - 2ae^{3x} = e^{3x}$$

giving $a = 1$, and a particular integral is $y = e^{3x}$, so that, adding on the complementary function found in the last example, we obtain the general solution as

$$y = Ae^{2x} + e^{3x}$$

3. $$y' - 2y = \sin 5x$$

(a) The complementary function is again

$$y = Ae^{3x}$$

(b) If we assume the form $y = a \sin 5x$ for the particular integral, we shall introduce a term in $\cos 5x$ arising from the differentiation of y. We therefore use the form

$$y = a \cos 5x + b \sin 5x$$

so that

$$y' = -5a \sin 5x + 5b \cos 5x$$

Substituting into the differential equation, we obtain

$$(-5a \sin 5x + 5b \cos 5x) - 2(a \cos 5x + b \sin 5x) = \sin 5x$$

or

$$(-5a - 2b) \sin 5x + (5b - 2a) \cos 5x = \sin 5x$$

This equation will be true for all values of x if we choose a and b so that the coefficients of $\sin 5x$ and $\cos 5x$ are the same on both sides. Equating these coefficients gives

$$-5a - 2b = 1 \quad \text{and} \quad 5b - 2a = 0$$

from which we find $a = -\frac{5}{29}$ and $b = -\frac{2}{29}$, so a particular integral is

$$y = -\tfrac{1}{29}(5 \cos 5x + 2 \sin 5x)$$

and the general solution is

$$y = A e^{2x} - \tfrac{1}{29}(5 \cos 5x + 2 \sin 5x)$$

These examples show clearly that the complementary function depends only on the left-hand side of the differential equation (that is, the terms containing y and y'), while the particular integral depends on the right-hand side (the terms with no y or y'). Table 9.1 relates the form of particular integral required to the right-hand side of the differential equation.

Table 9.1

Form of right-hand side	Form of particular integral
polynomial of degree n	polynomial of degree n
e.g. 1 ($n = 0$)	$y = a$
$1 + x - x^2$ ($n = 2$)	$y = ax^2 + bx + c$
x^3 ($n = 3$)	$y = ax^3 + bx^2 + cx + d$
e^{kx} (k constant)	$y = a e^{kx}$
$\sin kx$ or $\cos kx$	$y = a \cos kx + b \sin kx$
$e^{kx} \cos mx$ or $e^{kx} \sin mx$	$y = e^{kx}(a \cos mx + b \sin mx)$
$e^{kx} \times$ polynomial of degree n	$y = e^{kx} \times$ polynomial of degree n
$(\cos kx) \times$ polynomial of degree n	$y = (a \cos kx + b \sin kx) \times$ polynomial of degree n

The form of particular integral must be modified if it has a similar form to the complementary function, but we shall postpone discussion of this until we study second-order equations in Section 9.9.

EXERCISES 9.6.1

Find the complementary function, a particular integral and hence the general solution of the following differential equations:

(i) $y' - 3y = 3x$;   (ii) $y' - 3y = -6\sin 3x$;
(iii) $y' + 2y = e^{2x}$;   (iv) $y' + 2y = e^{2x}\cos 3x$;
(v) $y' + 2y = (1 + x^2)e^{3x}$.

9.7 SOLUTION OF SECOND-ORDER DIFFERENTIAL EQUATIONS

At the beginning of Section 9.1, we solved the differential equation $dv/dt = 1 - 2t$, with t representing time and v speed. Now replace v by ds/dt, where s is distance, so that

$$\frac{dv}{dt} = \frac{d}{dt}\left(\frac{ds}{dt}\right) = \frac{d^2s}{dt^2}$$

and we obtain the second-order differential equation

$$\frac{d^2s}{dt^2} = 1 - 2t \tag{1}$$

Integrating Equation (1) with respect to t gives

$$\frac{ds}{dt} = t - t^2 + c \tag{2}$$

Integrating again, we obtain

$$s = \tfrac{1}{2}t^2 - \tfrac{1}{3}t^3 + ct + d \tag{3}$$

Since we have performed two integrations to remove the second derivative, we have introduced two arbitrary constants. We call Equation (3) the *general solution* of the differential Equation (1). The general solution of any second-order differential equation must contain two arbitrary constants.

The initial condition we used before was $v(0) = 0$, but since $v = ds/dt$, this can be substituted into Equation (2) to give $c = 0$.

To determine the value of d, we need another initial condition, such as

$$s = 0 \text{ when } t = 0 \text{ (or } s(0) = 0)$$

Substituting this into Equation (3), we find $d = 0$, which gives

$$s = \tfrac{1}{2}t^2 - \tfrac{1}{3}t^3$$

Now consider the second-order differential equation

$$y'' - py' + qy = f(x) \tag{1}$$

in which p and q are constants and f is a function of x alone. This is not the most general form of second-order equation, because that would be too hard for us to tackle. It is a special case, called a *second-order linear differential equation with constant coefficients*. As usual, linear means that the degree of the terms in y and its derivatives is not greater than 1. Although the coefficients p and q are constant, the right-hand side is allowed to vary with x.

The method of solution copies that we used' latterly for first-order equations: that is, we split the solution into two parts:

(a) the complementary function, which is the general solution of the *homogeneous part* of Equation (1):

$$y'' - py' + qy = 0 \tag{2}$$

(b) a particular integral, which is any solution of Equation (1).

We concentrate for the moment on the homogeneous problem, Equation (2).

9.8 COMPLEMENTARY FUNCTION OF SECOND-ORDER DIFFERENTIAL EQUATIONS

We wish to find the general solution of the differential equation

$$y'' - py' + qy = 0 \tag{1}$$

Let us return for a moment to the first-order equation $y' - zy = 0$, where z is a constant. This is a separable (or linear) type and we easily see that its solution is $y = Ae^{zx}$. This suggests that we might try $y = e^{zx}$ as a solution of Equation (1). The resulting derivatives are $y' = ze^{zx}$ and $y'' = z^2 e^{zx}$. Putting these into Equation (1), we obtain

$$e^{zx}(z^2 - pz + q) = 0$$

This tells us that $y = e^{zx}$ is a solution of Equation (1) if z satisfies

$$z^2 - pz + q = 0 \tag{2}$$

since $e^{zx} \neq 0$. We call Equation (2) the *auxiliary equation*. It is a quadratic equation for z and, hence, has two roots, a and b say, so that, if these are distinct, there are two possible solutions of Equation (1),

$$y = e^{ax} \quad \text{and} \quad y = e^{bx}$$

We assert that

$$y = Ae^{ax} + Be^{bx} \qquad (3)$$

is the desired general solution, provided that $a \neq b$. For if this is the case, the second term cannot be a multiple of the first for all x, and so it contains the required two arbitrary constants. From Equation (3) we find

$$y' = Aae^{ax} + Bbe^{bx}$$

and

$$y'' = Aa^2e^{ax} + Bb^2e^{bx}$$

Substituting into the left-hand side of (1) gives

$$y'' - py' + qy = Aa^2e^{ax} + Bb^2e^{bx} - p(Aae^{ax} + Bbe^{bx}) + q(Ae^{ax} + Be^{bx})$$

$$= A(a^2 - pa + q)e^{ax} + B(b^2 - pb + q)e^{bx}$$

$$= 0$$

since a and b are roots of the auxiliary equation.

Examples 9.8.1

1.
$$y'' - 3y' + 2y = 0$$

Substituting $y = e^{zx}$, $y' = ze^{zx}$ and $y'' = z^2e^{zx}$ into the differential equation, we obtain

$$e^{zx}(z^2 - 3z + 2) = 0$$

so the auxiliary equation is

$$z^2 - 3z + 2 = 0$$

which has roots 1 and 2. The complementary function is therefore

$$y = Ae^{1x} + Be^{2x} = Ae^x + Be^{2x}$$

Note that the auxiliary equation can be written down directly from the differential equation, since it has the same coefficients. It is useful to recall the method of derivation, however, to obtain the correct form of solution when variables other than x and y are used.

2.
$$y'' - 4y' + 4y = 0 \qquad (1)$$

The auxiliary equation in this case is

$$z^2 - 4z + 4 = (z - 2)^2 = 0$$

so there is a repeated root of 2. It is no use writing the solution as

$$y = Ae^{2x} + Be^{2x}$$

293

since the second term is just a constant multiple of the first, and we can therefore add the two terms and replace $A + B$ by a single constant C. We need another solution which is not a multiple of this one. Such a solution is provided by $y = xe^{2x}$; we shall not derive this, but simply verify that it is a solution, since it is clearly not a constant multiple of the original solution. For if $y = xe^{2x}$ then $y' = (1 + 2x)e^{2x}$ and $y'' = 4(1 + x)e^{2x}$, and putting these into the left-hand side of Equation (1), we obtain

$$(4(1 + x) - 4(1 + 2x) + 4x)e^{2x}$$

which is zero, showing that $y = xe^{2x}$ is a solution of Equation (1). The general solution is now

$$y = (A + Bx)e^{2x}$$

3.
$$y'' - 6y' + 25y = 0 \qquad (1)$$

The auxiliary equation is

$$z^2 - 6z + 25 = 0$$

whose roots are $z = 3 \pm 4i$. These are certainly distinct, so we can write the solution as

$$y = Ae^{(3 + 4i)x} + Be^{(3 - 4i)x} \qquad (2)$$

It appears that the solution in this case is complex; however, let us put

$$e^{(3 \pm 4i)x} = e^{3x}e^{\pm 4ix} = e^{3x}(\cos 4x \pm i \sin 4x)$$

so that the solution becomes

$$y = e^{3x}(A(\cos 4x + i \sin 4x) + B(\cos 4x - i \sin 4x))$$
$$= e^{3x}(C \cos 4x + D \sin 4x)$$

where $C = A + B$ and $D = (A - B)i$ are new arbitrary constants. This makes it quite clear that, with real initial conditions, the solution is real for all values of x. The same initial conditions in Equation (2) would give values of A and B such that the terms combine to make the solution real.

We have now covered in examples all possible cases, which can be summarised as follows: When the roots of the auxiliary equation, a and b, are

(a) real and distinct, then the solution is
$$y = Ae^{ax} + Be^{bx}$$

(b) real and equal (to a, say), then the solution is
$$y = (A + Bx)e^{ax}$$

(c) complex ($e^{a \pm ib}$, say), then the solution is
$$y = e^{ax}(A \cos bx + B \sin bx)$$

EXERCISE 9.8.1

Solve the following differential equations:

(i) $y'' - 5y' + 6y = 0$; (ii) $y'' - y = 0$;
(iii) $y'' + 6y' + 9y = 0$; (iv) $y'' + 4y = 0$.

9.9 GENERAL SOLUTION OF NON-HOMOGENEOUS EQUATION

We write this in the form

$$y'' - py' + qy = f(x) \qquad (1)$$

for which the corresponding homogeneous equation is

$$y'' - py' + qy = 0 \qquad (2)$$

We show in the following Lemma and Theorem that the solution of Equation (1) is the sum of the complementary function, that is, the general solution of Equation (2), and a particular integral of Equation (1).

Lemma

If u and v are both solutions of Equation (1) then $u - v$ is a solution of Equation (2).

Proof Since u and v satisfy Equation (1), we have

$$u'' - pu' + qu = f(x)$$
$$v'' - pv' + qv = f(x)$$

Subtracting these equations, we find

$$(u - v)'' - p(u - v)' + q(u - v) = 0$$

which shows that $u - v$ is a solution of Equation (2), as required. $\square$

Theorem

If w is a particular integral of Equation (1) and u is the general solution of Equation (2), then the general solution of Equation (1) is $w + u$.

Proof Suppose that v is a solution of Equation (1). Then, by the Lemma, $v - w$ is a solution of Equation (2). But u is the general solution of Equation (2) and so we can write $v - w = u$, and, hence, $v = w + u$. Since v was arbitrary, this represents the general solution as required. $\square$

All that remains to be done is to find particular integrals for various forms of the function f in Equation (1). But we have already seen how to do this for first-order equations in Section 9.6; the technique is exactly the same for second-order equations, and the trial functions may be read off from Table 9.1.

Example 9.9.1

$$y'' + 3y' - 4y = 6x \qquad (1)$$

(a) The auxiliary equation is

$$z^2 + 3z - 4z = 0$$

which has roots 1 and -4. The complementary function is thus

$$y = Ae^x + Be^{-4x}.$$

(b) Since the right-hand side of Equation (1) is a linear function of x, we try the solution $y = ax + b$, which gives $y' = a$ and $y'' = 0$. Putting these into Equation (1), we find

$$0 + 3a - 4(ax + b) = 6x.$$

Since this must be true for all values of x, the terms in x and the constant terms must be the same on both sides, so we have $-4a = 6$ and $3a - 4b = 0$. These give $a = -\frac{3}{2}$ and $b = -\frac{9}{8}$, and so a particular integral is

$$y = -\tfrac{2}{3}x - \tfrac{9}{8}$$

and the general solution of Equation (1) is

$$y = -\tfrac{2}{3}x - \tfrac{9}{8} + Ae^x + Be^{-4x}.$$

We now give some examples of various combinations of left-hand sides and right-hand sides; they are not solved in detail, but the form of solution is indicated.

Examples 9.9.2

	Differential equation	Complementary function	Form of particular integral
1.	$y'' - y' - 6y =$ $e^{4x}\cos 2x$	$y = Ae^{3x} + Be^{-2x}$	$y = (a\cos 2x + b\sin 2x)e^{4x}$
2.	$y'' + 3y' - 10y =$ $(1 + 2x^2)e^{3x}$	$y = Ae^{2x} + Be^{-5x}$	$y = (a + bx + cx^2)e^{3x}$
3.	$y'' + 3y' - 10y =$ $x\sin 3x$	$y = Ae^{2x} + Be^{-5x}$	$y = (a + bx)\cos 3x$ $+ (c + dx)\sin 3x$

The above examples were carefully chosen so that the complementary function and the particular integral were dissimilar; we now show via an example how to treat cases when this is not so.

Example 9.9.3

$$y'' + 3y' - 4y = e^x \tag{1}$$

(a) Since the homogeneous part is the same as in Example 9.9.1, the complementary function is also the same.

(b) For a particular integral we try $y = ae^x$. Putting this into Equation (1), we find

$$a(e^x + 3e^x - 4e^x) = e^x$$

which is impossible, since the left-hand side vanishes. Of course, the reason for this is that e^x is part of the complementary function, which must cause the left-hand side to vanish, since it is a solution of the corresponding homogeneous equation.

We use exactly the same device here that we used to obtain a new solution when the auxiliary equation had repeated roots, that is, we multiply by x. Our trial solution is now $y = axe^x$, giving $y' = a(1 + x)e^x$ and $y'' = a(2 + x)e^x$. Putting these into Equation (1), we find

$$((2 + x) + 3(1 + x) - 4x)ae^x = e^x$$

The terms in x vanish, while equating the other terms on both sides of this equation gives $a = 1$. Thus, a particular integral is $y = xe^x$ and the general solution is

$$y = (A + x)e^x + Be^{-4x}$$

Generally, if the trial solution for a particular integral suggested by the form of the right-hand side contains terms which are also in the complementary function, then the trial solution should be multiplied by x. A briefer way of saying this, which we shall adopt, is 'if the complementary function and particular integral overlap, then multiply the latter by x'. When the auxiliary equation has a repeated root, the complementary function will already have a term multiplied by x; in this case the trial solution needs multiplying by x^2. We give below some examples where this arises, contenting ourselves with suggesting suitable forms of particular integral but omitting detailed solutions.

297

Examples 9.9.4

1.
$$y'' - 6y' + 13y = 7e^{3x}\sin 2x$$

The complementary function is

$$y = e^{3x}(A\cos 2x + B\sin 2x)$$

Our first suggestion for a particular integral (from Table 9.1) would be

$$y = e^{3x}(a\cos 2x + b\sin 2x)$$

but since this is the same as the complementary function, we must multiply by x to obtain

$$y = xe^{3x}(a\cos 2x + b\sin 2x)$$

2.
$$y'' - y' - 6y = (1 + 2x^2)e^{3x}$$

Here the complementary function is $y = Ae^{3x} + Be^{-2x}$, while Table 9.1 suggests a particular integral of the form

$$y = (a + bx + cx^2)e^{3x}$$

These overlap, since each contains a term of the form constant $\times e^{3x}$, so we take

$$y = (a + bx + cx^2)xe^{3x}$$

for the form of particular integral.

3.
$$y'' - 6y' + 9y = (1 + 2x^2)e^{3x}$$

The complementary function is $y = (A + Bx)e^{3x}$, which would give a 'double' overlap with the trial particular integral, so we multiply by x^2 to obtain for the form of particular integral

$$y = (a + bx + cx^2)x^2e^{3x}$$

4.
$$y'' + 4y = e^{4x}\cos 2x$$

The complementary function is $y = A\cos 2x + B\sin 2x$, while the form of particular integral suggested by Table 9.1 is

$$y = (a\cos 2x + b\sin 2x)e^{4x}$$

Although the two parts of the solution look similar, they are not the same because of the multiplier e^{4x} in the latter, so we must not multiply by x here.

These examples show that we should always find the complementary function first, so that we can check our trial form of particular integral for overlap. It is important to make sure that the overlap is genuine, and not apparent as in Example 9.9.4.4.

EXERCISES 9.9.1

1 Find the general solution of the differential equation $y'' - 5y' + 6y = f(x)$ for the following cases:

(i) $f(x) = x^2$; (ii) $f(x) = 2e^x$; (iii) $f(x) = e^{2x}$;
(iv) $f(x) = \cos 7x + 2 \sin 7x$.

2 Find the general solution of the differential equation $y'' + 2y' + y = e^{-x}$.

3 Suggest suitable forms of particular integral (but do not evaluate the constants) for the following differential equations:

(i) $y'' + 9y = x \sin 3x$;
(ii) $y'' - y' - 6y = (\sin 5x - 2 \cos 5x)e^{3x}$;
(iii) $y'' - 2y' + 26y = xe^x \cos 5x + (1 - x^2)e^x \sin 5x$;
(iv) $y'' - 6y' + 13y = e^{4x} \sin 2x$.

9.10 INITIAL AND BOUNDARY CONDITIONS

We have already seen that two initial conditions are required for a second-order differential equation. These would normally consist of values of y and y' prescribed for a given value of x. An alternative is to specify values of y for two different values of x; we call these *boundary conditions*. Either sort of conditions leads to two simultaneous equations for finding the constants.

Example 9.10.1

For Example 9.9.3, suppose that the initial conditions are $y = 4$ and $y' = 0$ when $x = 0$.

The general solution is

$$y = (A + x)e^x + Be^{-4x}$$

so the first condition gives $A + B = 4$. Also

$$y' = (A + 1 + x)e^x - 4Be^{-4x}$$

and the second condition gives $A + 1 - 4B = 0$. These equations are satisfied by $A = 3$ and $B = 1$, so the required solution is

$$y = (3 + x)e^x + e^{-4x}$$

EXERCISES 9.10.1

1 Find the solution of the differential equation $2y'' - 2y' = -5$ for which $y(0) = 1$ and $y'(0) = 4$.

2 Solve the differential equation $y'' + 4y = 0$, where $y = 1$ when $x = 0$ and $y = 0$ when $x = \pi/4$.

MISCELLANEOUS EXERCISES 9

1 In an adiabatic expansion of a gas, the pressure p and volume v satisfy the differential equation

$$\frac{dp}{dv} = -\gamma \frac{p}{v},$$

where γ is a constant. Show that $pv^\gamma = \text{constant}$.

2 The rate of fall of temperature θ of a body is determined by Newton's law of cooling,

$$-\frac{d\theta}{dt} = k(\theta - \theta_0),$$

where θ_0 is the temperature of the surroundings and k is a constant. If $\theta = \theta_1$ when $t = 0$, find the temperature of the body at time t.

3 The current i caused by a voltage V applied to a circuit of resistance R and inductance L is determined by the differential equation

$$L\frac{di}{dt} + Ri = V$$

Find the general solution assuming that L, R, V are constant and show that i approaches a constant value (the *terminal current*) at large times.

4 An animal population P obeys the *logistic equation*

$$\frac{dP}{dt} = aP\left(1 - \frac{P}{M}\right)$$

where a, M are positive constants. Find and describe the nature of the solution for which $P = P_0$ when $t = 0$, where $0 < P_0 < M$.

5 In a particular second-order chemical reaction the amount x of a product at time t satisfies the differential equation

$$\frac{dx}{dt} = k(1 - x)(3 - x)$$

where k is a constant. Find the solution for which $x = 0$ when $t = 0$, and hence find the time taken for the amount of the product to reach x.

6 The position x of a body of mass m acted upon by a force $F \cos \omega t$ satisfies the differential equation $m \dfrac{d^2 x}{dt^2} = F \cos \omega t$, where ω is a constant.

If the body starts from rest at $t = 0$, find how far it has moved after time t.

7 A sphere of mass m falling from rest under the effect of gravity and with air resistance F has speed v which satisfies the differential equation $\dfrac{dv}{dt} = g - \dfrac{F}{m}$, where g is the acceleration due to gravity. Solve this equation for v in the cases when

  (i) $F = kv$
  (ii) $F = kv^2$

where k is a constant. Show that at large times the speed becomes almost a constant (the *terminal speed*), in the first case $\dfrac{mg}{k}$ and in the second $\sqrt{\dfrac{mg}{k}}$.

8 Find the general solution of the differential equation

$$\frac{d^2 y}{dx^2} + 2\alpha \frac{dy}{dx} + \omega^2 y = 0$$

Show that

  (i) if $\alpha^2 - \omega^2 > 0$, then the solution is a decreasing exponential if $\alpha > 0$ and an increasing exponential if $\alpha < 0$;
  (ii) if $\alpha^2 - \omega^2 < 0$, then the solution oscillates with decreasing amplitude if $\alpha > 0$ and with increasing amplitude if $\alpha < 0$.

9 The displacement x of a mass attached to a fixed point by a spring of natural length l_0 and stiffness k satisfies the differential equation

$$m \frac{d^2 x}{dt^2} = k(l_0 - x)$$

Show that the general solution may be written

$$x(t) = A \cos \omega t + B \sin \omega t + l_0$$

where $\omega = \sqrt{k/m}$. Find the values of A and B if the mass is released from rest with the spring stretched to a length x_0.

10 For small displacements q of the bob of a simple pendulum of length l the differential equation $l \dfrac{d^2 q}{dt^2} + gq = 0$ is approximately satisfied, where g is the acceleration due to gravity. Obtain the solution in the form $q = A \cos \omega t + B \sin \omega t$, where $\omega = \sqrt{g/l}$.

9.11 ANSWERS TO EXERCISES

Exercise 9.1.1

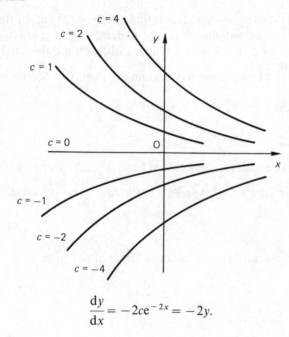

$$\frac{dy}{dx} = -2ce^{-2x} = -2y.$$

Exercises 9.2.1

1 (i) $y = \int \cos x \, dx = \sin x + c,$

(ii) $y = 2 \int \frac{dx}{x} = 2 \ln |kx|,$

(iii) $y = -\int \left(1 + \frac{1}{x}\right) dx = -x - \ln |x| + c$ (or $-x - \ln |kx|$).

2 (i) $\int e^y \, dy = x + c \Rightarrow e^y = x + c \Rightarrow y = \ln |x + c|,$

(ii) $\int 2y \, dy = x + c \Rightarrow y^2 = x + c,$

(iii) $\int \left(\frac{1}{y} + 1\right) dy = x + c \Rightarrow \ln |y| + y = x + c,$

(iv) $\int 2x \, dx = t + c \Rightarrow x^2 = t + c.$

3 (i) $\displaystyle\int y^{-1}\,dy = \int x\,dx \Rightarrow \ln|ky| = \frac{1}{2}x^2 \Rightarrow y = k'e^{x^2/2}$, where $k' = k^{-1}$,

(ii) $\displaystyle\int y^{-2}\,dy = \int \sin x\,dx \Rightarrow -y^{-1} = -\cos x + c \Rightarrow y = \frac{1}{\cos x + c}$,

(iii) $\displaystyle\int y^{-1}\,dy = \int (x+1)^{-1}\,dx \Rightarrow \ln|y| = \ln|k(x+1)| \Rightarrow y = k(x+1)$,

(iv) $\displaystyle\int y^{-1}\,dy = \int \frac{x+2}{x(x+1)}\,dx \Rightarrow \ln|y| = \int \left(\frac{2}{x} - \frac{1}{x+1}\right) dx$

$$= 2\ln|x| - \ln|x+1| + \ln|k| = \ln\left|\frac{kx^2}{x+1}\right| \Rightarrow y = \frac{kx^2}{x+1}.$$

Exercise 9.3.1

The substitution $y = vx$, $y' = v + v'x$ gives

$$v + v'x = 1 + v + v^2 \Rightarrow v'x = 1 + v^2 \Rightarrow \int \frac{dv}{1+v^2} = \int \frac{dx}{x}$$

$$\Rightarrow \tan^{-1} v = \ln|kx| \Rightarrow v = \tan\ln|kx| \Rightarrow y = x\tan\ln|kx|.$$

Exercise 9.4.1

(i) Rearrange the differential equation to $3x^2 - y + y'(2y - x) = 0$. Since $\dfrac{\partial}{\partial y}(3x^2 - y) = -1 = \dfrac{\partial}{\partial x}(2y - x)$, it is exact; in fact, it can be written $\dfrac{d}{dx}(x^3 - xy + y^2) = 0$, so the solution is $x^3 - xy + y^2 = c$.

(ii) $\dfrac{\partial}{\partial x}\cos x = -\sin x = \dfrac{\partial}{\partial y}(-y\sin x)$, hence exact. It may be written $\dfrac{d}{dx}(y\cos x) = 0$, which integrates to $y\cos x = c$ or $y = c\sec x$.

Exercises 9.5.1

1 (i) The integrating factor is e^{2x}, so

$$y'e^{2x} + 2ye^{2x} = e^{2x} \Rightarrow \frac{d}{dx}(ye^{2x}) = e^{2x} \Rightarrow ye^{2x} = \frac{1}{2}e^{2x} + c \Rightarrow y = \frac{1}{2} + ce^{-2x};$$

(ii) the integrating factor is $e^{-2\int(dx/x)} = e^{-2\ln x} = e^{\ln x^{-2}} = x^{-2}$, so

$$\frac{d}{dx}(yx^{-2}) = x^2 \cdot x^{-2} = 1 \Rightarrow yx^{-2} = x + c \Rightarrow y = x^2(x + c);$$

(iii) the integrating factor is $e^{\int (dx/x)} = e^{\ln x} = x$, so

$$\frac{d}{dx}(yx) = \frac{1}{x} \cdot x = 1 \Rightarrow yx = x + c \Rightarrow y = 1 + \frac{c}{x};$$

(iv) the integrating factor is $e^{-\int 2x \, dx} = e^{-x^2}$, so

$$\frac{d}{dx}(ye^{-x^2}) = e^{x^2} \cdot e^{-x^2} = 1 \Rightarrow ye^{-x^2} = x + c \Rightarrow y = (x + c)e^{x^2};$$

(v) rearranging to standard form gives $y' - y = 1$, so the integrating factor is $e^{-\int dx} = e^{-x}$, so

$$\frac{d}{dx}(ye^{-x}) = e^{-x} \Rightarrow ye^{-x} = -e^{-x} + c \Rightarrow y = 1 + ce^x.$$

2 (i) Linear, (ii) exact, (iii) separable $\left(\dfrac{y'}{y - 2y^2} = \dfrac{1}{x}\right)$, (iv) separable, (v) separable.

Exercises 9.6.1

(i) Complementary function Ae^{3x}, particular integral $y = ax + b$ with $a = -1$, $b = -\frac{1}{3}$, general solution $y = Ae^{3x} - x - \frac{1}{3}$.

(ii) Complementary function Ae^{3x}, particular integral $y = a \cos 3x + b \sin 3x$ with $a = b = 1$, general solution $y = Ae^{3x} + \cos 3x + \sin 3x$.

(iii) Complementary function Ae^{-2x}, particular integral $y = ae^{2x}$ with $a = \frac{1}{4}$, general solution $y = Ae^{-2x} + \frac{1}{4}e^{2x}$.

(iv) Complementary function Ae^{-2x}, particular integral $y = e^{2x}(a \cos 3x + b \sin 3x)$ with $a = \frac{4}{25}$, $b = \frac{3}{25}$, general solution

$$y = Ae^{-2x} + \tfrac{1}{25}e^{2x}(4 \cos 3x + 3 \sin 3x).$$

(v) Complementary function Ae^{-2x}, particular integral $y = (a + bx + cx^2)e^{3x}$ with $a = \frac{27}{125}$, $b = -\frac{2}{25}$, $c = \frac{1}{5}$, general solution

$$y = Ae^{-2x} + \tfrac{1}{125}(27 - 10x + 25x^2)e^{3x}.$$

Exercise 9.8.1

(i) $y = Ae^{2x} + Be^{3x}$,

(ii) $y = Ae^x + Be^{-x}$,

(iii) $y = (A + Bx)e^{-3x}$,

(iv) $y = Ae^{2ix} + Be^{-2ix}$ or $y = C \cos 2x + D \sin 2x$.

Exercises 9.9.1

1 The complementary function for all cases is $y = Ae^{2x} + Be^{3x}$. This should be added to the following particular integrals to obtain the general solution for each case:

  (i) $y = ax^2 + bx + c$ is a particular integral with $a = \frac{1}{6}, b = \frac{5}{18}, c = -\frac{19}{108}$,

  (ii) $y = ae^x$ is a particular integral with $a = 1$.

  (iii) $y = axe^{2x}$ is a particular integral (note that the factor x is included to avoid overlap with the complementary function) with $a = -1$,

  (iv) $y = a\cos 7x + b\sin 7x$ is a particular integral with $a = \frac{27}{3074}, b = -\frac{121}{3074}$.

2 The complementary function is $y = (A + Bx)e^{-x}$ (note the coincident roots of the auxiliary equation). To avoid overlap, we use $y = ax^2e^{-x}$ for the trial solution and find $a = \frac{1}{2}$.

3 (i) $y = (ax + bx^2)\cos 3x + (cx + dx^2)\sin 3x$ (note overlap),

  (ii) $y = (a\cos 5x + b\sin 5x)e^{3x}$,

  (iii) $y = e^x\{(ax + bx^2 + cx^3)\cos 5x + (dx + fx^2 + gx^3)\sin 5x\}$ (overlap here),

  (iv) $y = e^{4x}(a\cos 2x + b\sin 2x)$ (no overlap).

Exercises 9.10.1

1 The general solution is $y = A + Be^x + \frac{5}{2}x$. $y(0) = 1$ gives $A + B = 1$ and $y'(0) = 4$ gives $B + \frac{5}{2} = 4$, so $A = -\frac{1}{2}$, $B = \frac{3}{2}$ and the solution is $y = -\frac{1}{2} + \frac{3}{2}e^x + \frac{5}{2}x$.

2 The general solution is $y = A\cos 2x + B\sin 2x$. $y(0) = 1$ gives $A = 1$ and $y\left(\frac{\pi}{4}\right) = 0$ gives $B = 0$, so the solution is $y = \cos 2x$.

Miscellaneous Exercises 9

1 $\displaystyle \int \frac{dp}{p} = -\gamma \int \frac{dv}{v} \Rightarrow \ln|p| = -\gamma \ln|v| + C \Rightarrow \ln|pv^\gamma| = C.$

2 $\displaystyle \int \frac{d\theta}{(\theta - \theta_0)} = -k \int dt \Rightarrow \ln|\theta - \theta_0| = -kt + \ln|c| \Rightarrow \theta - \theta_0 = ce^{-kt}.$

  $\theta = \theta_1$ when $T = 0$ gives $c = \theta_1 - \theta_0$.

3 $\dfrac{di}{dt} + \dfrac{R}{L}i = \dfrac{V}{L}$ is linear with integrating factor $e^{Rt/L}$, so

$$\frac{d}{dt}(ie^{Rt/L}) = \frac{V}{L}e^{Rt/L} \Rightarrow ie^{Rt/L} = \frac{V}{R}e^{Rt/L} + C \Rightarrow i = \frac{V}{R} + Ce^{-Rt/L}.$$

As $t \to \infty$, $i \to \dfrac{V}{R}$.

4
$$\int \frac{M\,dP}{P(P-M)} = -a\int dt \Rightarrow \int\left(\frac{1}{P-M} - \frac{1}{P}\right)dP = -a\int dt$$

$$\Rightarrow \ln\left|\frac{P-M}{PC}\right| = -at \Rightarrow \frac{P-M}{P} = Ce^{-at}.$$

The initial condition gives

$$C = \frac{P_0 - M}{P_0}$$

and solving for P gives $P = \dfrac{MP_0}{P_0 + (M-P_0)e^{-at}} \to M$ as $t \to \infty$. This is the so-called *logistic* curve shown below.

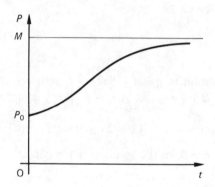

5 $\displaystyle\int \frac{dx}{(x-1)(x-3)} = k\int dt$. The solution is as in Miscellaneous Exercise 3.6:

$$x = \frac{e^{2kt} - 1}{e^{2kt} - 1/3} \text{ if } x < 1 \text{ or } x > 3 \text{ giving } t = \frac{1}{2k}\ln\left(\frac{x/3 - 1}{x - 1}\right),$$

$$x = \frac{e^{2kt} + 1}{e^{2kt} + 1/3} \text{ if } 1 < x < 3 \text{ giving } t = \frac{1}{2k}\ln\left(\frac{1 - x/3}{x - 1}\right).$$

6 Integrating with respect to t, $m\dfrac{dx}{dt} = F\displaystyle\int \cos \omega t\, dt = \frac{F}{\omega}\sin \omega t + c$. $\dfrac{dx}{dt} = 0$

when $t = 0 \Rightarrow c = 0$. Integrating again, $mx = -\dfrac{F}{\omega^2}\cos \omega t + d$. Putting

$x = 0$ when $t = 0$ gives $d = \dfrac{F}{\omega^2}$, so $x = \dfrac{F}{m\omega^2}(1 - \cos \omega t)$.

7 (i) $\displaystyle\int\frac{dv}{g-kv/m}=\int dt\Rightarrow-\frac{m}{k}\ln|g-kv/m|=t-\frac{m}{k}\ln|c|$

$$\Rightarrow g-kv/m=ce^{-kt/m}\to0\text{ as }t\to\infty,\text{ so }v\to\frac{mg}{k}.$$

(ii) $\displaystyle\int\frac{dv}{g-kv^2/m}=\int dt\Rightarrow\int\frac{dv}{gm/k-v^2}=\frac{k}{m}t+\ln|c|$

$$\Rightarrow\frac{1}{2\sqrt{gm/k}}\int\left(\frac{1}{\sqrt{gm/k}+v}+\frac{1}{\sqrt{gm/k}-v}\right)dv$$

$$=\frac{k}{m}t+\ln|c|\Rightarrow\frac{1}{2\sqrt{gm/k}}\ln\left|\frac{\sqrt{gm/k}+v}{\sqrt{gm/k}-v}\right|=\frac{k}{m}t+\ln|c|.$$

As $t\to\infty$ it is clear that $v\to\sqrt{\dfrac{gm}{k}}$.

8 The auxiliary equation is $\lambda^2+2\alpha\lambda+\omega^2=0$ giving $\lambda=-\alpha\pm\sqrt{\alpha^2-\omega^2}$ and the general solution is $y=Ae^{(-\alpha+\sqrt{\alpha^2-\omega^2})x}+Be^{(-\alpha-\sqrt{\alpha^2-\omega^2})x}$.

(i) $\alpha^2-\omega^2>0\Rightarrow$ real roots,

$$-\alpha+\sqrt{\alpha^2+\omega^2}<0\quad\text{and}\quad-\alpha-\sqrt{\alpha^2-\omega^2}<0\quad\text{if}\quad\alpha>0,$$

hence decreasing exponential:
$-\alpha+\sqrt{\alpha^2-\omega^2}>0$ and $-\alpha-\sqrt{\alpha^2-\omega^3}>0$ if $\alpha<0$, hence increasing exponential;

(ii) $\alpha^2-\omega^2<0\Rightarrow\lambda$ complex, so $y=e^{-\alpha x}(C\cos x\sqrt{\omega^2-\alpha^2}+D\sin x\sqrt{\omega^2-\alpha^2})$.

9 $\dfrac{d^2x}{dt^2}+\dfrac{k}{m}x=0$ has solution $x=A\cos t\sqrt{\dfrac{k}{m}}+B\sin t\sqrt{\dfrac{k}{m}}$. A particular integral is $x=l_0$, so the general solution is $x=A\cos\omega t+B\sin\omega t+l_0$, where $\omega=\sqrt{\dfrac{k}{m}}$. Putting $x=x_0$ when $t=0$ gives $A=x_0-l_0$ and $\dfrac{dx}{dt}=0$ when $t=0$ gives $B=0$, so the required solution is $x=(x-l_0)\cos\omega t+l_0$.

10 $\dfrac{d^2q}{dt^2}+\omega^2q=0$ has auxiliary equation $\lambda^2+\omega^2=0$, so $\lambda=\pm i\omega$ and the solution is $q=A\cos\omega t+B\sin\omega t$.

INDEX

arbitrary constant 53
arclength 162–163
Argand diagram 248

cartesian coordinate system 132
chain rule 183–184
completing the square 75
complex numbers 247–262
  argument of 258
  conjugate of 251
  exponential form of 261–262
  modulus of 251
  polar form of 257
continuity 6–8
convergence
  interval of 40
  radius of 40
cross product, *see* vector product
cylinders 178

de Moivre's Theorem 260
derivatives 28
  of combination of
     functions 29–33
  of exponential functions 36–37
  of function of functions 32
  higher 38–39
  of hyperbolic functions 37
  of inverse functions 33
  of logarithm functions 36–37
  of trigonometric functions 33–35
determinants 98, 100–105
  of coefficients 98

cofactor of 100
element of 98
expansion of 100
minor of 100
order of 98
differentiable 28
differential equations
  first order 270–291
    complementary function of 288
    exact 282–283
    family of solution curves of 272
    general solution of 274
    initial condition for 274
    linear 283–286
    order of 274
    particular integral of 288–291
    of separable type 275–280
  second order 291–299
    boundary condition for 299
    complementary function
      of 292–295
    general solution of 295–297
    initial condition for 299
    parameter of 272
differential operator 28
differentiation 26–43
  chain rule for 32
  from first principles 28
  product rule for 30
  quotient rule for 30
direction cosine 138
directional derivative 185–186
discontinuity 6

discriminant 192
distance
  of plane from origin 153
  of point from line 150
dot product, see scalar product
double integral 215–236
  change of variable in 226–233

fields
  conservative 209–215
  scalar 202
  vector 202
functions 1–19
  of one variable 4–6
    composition of 15–19
    continuous 8
    defined by 5
    derivative of 28
    differentiable 28
    domain of 5
    exponential 11–13
    family of standard functions 6,
      19
    hyperbolic 14–15
    inverse 15–19
    linear 58
    log, logarithm 11–13
    principal value of 18
    range of 5
    rational 9
    restriction of 17
    trigonometric 9–11
    value of 5
  of two variables 175–193
    cross-section of 178
    level curve of 177
    maximum and minimum of 190
    saddle point of 191
    standard family of 176
    stationary point of 190
  of three or more variables 181
Fundamental Theorem of
    Calculus 52–53

Gaussian elimination 121–123
  back-substitution 121
  maximum pivoting 122
  pivot 122
Green's Theorem 234–236

implicit differentiation 37
infinity 7

integrals
  definite 49
  improper 78–81
    infinite integrand 80–81
    infinite range 78–80
  indefinite 53
  list of the standard 81
integrand 49
integration 48–84
  by parts 64–69
  by reduction formula 66
  by substitution 55–64
  interval of 49
  limits of 49
  of rational functions 73–76
  of rational trigonometric
      functions 77–78
interval 1
irreducible quadratic 74

Jacobian 226

Leibniz's Rule 39
limit 6–8
line integrals 202–215
linear equations 96–124
lines and planes 149–155
long division 70

matrices 106–112
  augmented 115
  of coefficients 106
  elementary row operations on 116
  identity 112
  inverse 112–118
  null 111
  numerical solution of 121–124
  singular 114
  square 112–118
  transpose 119

orthogonal projection 152
Osborne's Rule 15

parametric equations of curve 157
  of plane 153
partial derivatives 181–185
  chain rule for 183–184
  higher 186–189
  operator form of 187
partial fractions 69–71
polynomials 8
  factorisation of 70

position vectors   134
potential   210
power series   40–41
  expansion   40
primitive   53

quadratic equations   253

real line   3
repeated integrals   216–220
right-handed triad   131

saddle   179
scalar   131
scalar products   140–141
sequence   1
sets   1–3
simply connected region   212
simultaneous equations   96
  homogeneous   96
  inconsistent   97
  infinity of solutions of   97
  trivial solution of   96
  unique solution of   105
singular point of transformation   227

tangent vectors   160
tends to   7
trigonometric identities   263–264
triple scalar products   143–144
triple vector products   147

unique   4

vector equations
  of curve in space   156–158
  of line   149
  of plane   152
vector products   142–146
vectors   107, 131–163
  component of   107, 134
  derivative of   159
  gradient of   203
  magnitude of   134
  notation for   134
  parallelogram law for addition
    of   134
  sense of   135
  triangle law for addition of   134
  unit   137
  zero   137
vector-valued functions   156